P9-BZH-363

Are Dolphins Really Smart?

JUSTIN GREGG

OXFORD
UNIVERSITY PRESS

Great Clarendon Street, Oxford, OX2 6DP,
United Kingdom

Oxford University Press is a department of the University of Oxford.
It furthers the University's objective of excellence in research, scholarship,
and education by publishing worldwide. Oxford is a registered trade mark of
Oxford University Press in the UK and in certain other countries

First Edition published in 2013
Impression: 1

British Library Cataloguing in Publication Data
Data available

ISBN 978–0–19–966045–2

Printed in Great Britain by
CPI Group (UK) Ltd, Croydon, CR0 4YY

ACKNOWLEDGMENTS

I would like to thank the small army of colleagues, experts, and dolphin aficionados whom I consulted while writing this book. I must apologize for all the advice they've given that I willfully ignored, completely misunderstood, or otherwise misinterpreted. I am entirely responsible for any and all flawed logic or wonky discussion points. The list of those I wish to thank, in random order, is Lou Herman, Stefan Huggenberger, David Janiger, Joan E. Roughgarden, D. Graham Burnett, Manuel Garcia Hartmann, Achim Winkler, Jeanette Thomas, Tengiz Zorikov, Mikhail Ivanov, Mike Johnson, Paul Nachtigall, Suzana Herculano-Houzel, Brenda McCowan, Camilla Butti, Tadamichi Morisaka, Rebecca Singer, Fabienne Delfour, Mike Johnson, and Jonathan Balcombe.

A special thanks goes out to friends and colleagues who read initial drafts of some sections of the book, and provided fantastically helpful feedback and suggestions, including Stan Kuczaj, Patrick Hof, Lori Marino, and Sabrina Brando. A special thanks to Thomas White for a productive and enjoyable email exchange.

I am eternally grateful to my colleagues and friends at the Dolphin Communication Project; Kathleen Dudzinski and Kelly Melillo-Sweeting. Kathleen and Kelly have been with me every step of the way during the process of writing this book, and provided invaluable feedback, advice, and encouragement. Thanks also to Howard Smith for giving me the chance to pursue research into dolphin cognition, and setting me on the path that led to the writing of this book.

I wish to thank the publishing team at Oxford University Press for providing me the opportunity to present this information to the public. Latha Menon has been an outstanding editor, providing me much needed guidance and invaluable advice along the way.

I also wish to thank my colleagues at *Aquatic Mammals* and Document and Publication Services, including Sandy Larimer, Tammy Carson, Laura Caldwell, and Gina Colley.

I have sought the support of many friends during the writing of this book, and I would like to thank those who have helped me along the way with advice, encouragement, and lack of protest when I asked them to read section of the book, including Genevieve Bergendahl, Peter Bergendahl, Stacy Mosel, Erin Cole, Tim Ecott, Mark Westbrook, Brendan Lucey, Loredana Loy, Sunna Edberg, and Sandro Magliocco.

A very special thank you goes out to my parents, David and Susan Gregg. They have always supported me in everything I have ever wanted to do; a rare and incredibly generous quality. Any success I have in life will ultimately finds its roots in the many years of toil and trouble that they have undertaken on my behalf. Thank you!

And last and far from least, a special thank you to my wife Ranke de Vries, and my daughter Mila. I'm afraid that Ranke has been subjected to more monologues about dolphin cognition than is advisable for maintaining the health of a marriage, and yet she's still there every morning to share a cup of coffee and continue to formulate our madcap plans and future adventures. And Mila continues to provide me with the happiest moments in my life. Even if this book should sell a billion copies, Mila will remain the most amazing and successful thing I have ever produced.

CONTENTS

1 The Second Most Intelligent Creature on Earth

The fact that an opinion has been widely held is no evidence whatever that it is not utterly absurd; indeed in view of the silliness of the majority of mankind, a widespread belief is more likely to be foolish than sensible.[1] *Bertrand Russell*

This is the story of a truly remarkable animal. Its bond with humankind goes back many centuries, if not millennia, and it appears in our earliest myths and legends. Recent scientific research has revealed its sophisticated social behavior and remarkable cognitive abilities. With these discoveries, many who once spoke of the undeniable superiority of the human intellect have been humbled into silence. What's more, this extraordinary creature is only distantly related to humans and other primates, which makes the following discoveries of its intellectual prowess all the more startling:

- They can live in groups numbering in the hundreds and are able to recognize and remember individuals within the group, as well as their place in a complex social hierarchy.[2]
- Certain behaviors appear to be learned from other members in the group: a form of social transmission of behavior traditionally

seen only in primates, and the first steps toward what scientists now refer to as animal culture.[3]

- They have been shown to display signs of empathy—registering increased heart rate and anxious behaviors when watching their friends or family in distress.[4]
- When food is found, they will alert members of the group by producing a complex series of vocalizations that vary depending on the kind of food. What's more, these vocalizations appear to refer to food in a one-to-one correlation—almost like human words.[5]
- They produce unique vocalizations that appear to refer to different threats in the environment—these vocalizations are meant to warn the group of approaching danger. Individuals will take evasive action that differs depending on the kind of vocalization they hear.[6]
- In experimental conditions, they have been shown to have the ability to anticipate future events, and to exercise self-control—delaying an immediate food reward when they realize that by waiting they may get an even larger food reward. Thus, like humans, they may be able to plan for the future.[7]

This inventory of astonishingly complex cognitive talents belongs not to the dolphin (as one might have expected based on the title of this book), but to the humble domestic chicken. Surprised? If so, you can take some comfort in the fact that some (although not all) of these complex behaviors have also been observed in dolphins. The lesson here, however, is that chickens are not as dim-witted as popular opinion would have us believe. Long-held ideas of animal intelligence are increasingly out of step with the rapid pace with which the vanguard of science penetrates the mysteries of the animal mind. If Bertrand Russell's tongue-in-cheek maxim as to the silliness of mankind is as insightful as it is funny, there's good reason to subject these popular ideas to careful scrutiny.

Popular opinion considers dolphins to be fairly intelligent animals. As we shall see, there is ample scientific evidence to lend support to this idea. But if we're going to think critically about what it means to be "a fairly intelligent animal," we'll need to do three things: 1) produce a working definition of *intelligence*; 2) get to grips with what the scientific literature on dolphin intelligence is (or is not) telling us; and 3) put these findings in proper context by examining what is known about the intelligence of other species. We'll need to ponder the implications of this list of ostensibly complex chicken behavior, and decide what it means that the veined octopus, with a brain smaller than a dolphin's eyeball, is known to use tools with a sophistication rivaling that of dolphins.[8] The past few decades have produced many startling discoveries of complex animal behavior scattered throughout the taxa, including in animals previously considered unintelligent. Science continues to challenge our traditional ideas about where individual species should be placed on the continuum of animal intelligence, and whether a continuum of animal intelligence is a useful model to begin with. Dolphins might well be smart, but the idea that they sit atop a pedestal as "the second most intelligent species on the planet"[9] is now passé. Under Ben Goldacre's banner of "I think you'll find it's a bit more complicated than that,"[10] I can safely state that the notion that *dolphins are smart and chickens are stupid* is at best a gross oversimplification, and at worst completely wrong and thoroughly unhelpful.

The Myth of the Intelligent Dolphin

The myth of the intelligent dolphin has been percolating in popular culture for more than half a century, and comprises a number of beliefs about dolphins, ranging from the innocuous (e.g., dolphin societies are unusually complex) to the bizarre (e.g., dolphins can teleport people to Mars).[11] I have distilled the myth into five core themes that will be carefully dissected in this book:

1. The idea that the dolphin brain is unusually large and sophisticated.
2. The idea that the dolphin mind is unusually complex when it comes to self-awareness, consciousness, and emotions.
3. The idea that dolphins display unusually sophisticated behavior in the wild and in experimental situations.
4. The idea that dolphins speak dolphinese, a vocal communication system as complex as human language, which scientists will one day decipher.
5. The idea that dolphins lead usually complex social lives, and live in peaceful harmony with each other and their environment.

My aim in writing this book is to deconstruct the dolphin myth, and to determine to what extent the ideas underpinning these five themes are based on scientific fact as opposed to science fiction. By remaining impartial, skeptical, and critical of the results of scientific experiments and observations of dolphin behavior, it is possible to create a realistic picture of these much-loved animals. I won't be able to tackle every claim about dolphins that can be found in popular culture, as there simply isn't enough space in this book (or any book that is expected to be bound in a single volume). There are probably more weird ideas about dolphins swimming in cyberspace than there are dolphins swimming in the ocean. Assuming that you, kind reader, are sympathetic to the concepts of critical thinking, skepticism, and the value of science at explaining the natural world, there is no need for me to go into detail as to why dolphins most likely cannot teleport you to Mars.

The Legend of Lilly

As I began doing research for this book and tracing the history of the development of the many ideas about dolphins we see in popular culture today, I was surprised to learn that nearly all of the principal concepts underpinning the five major themes can be traced to a single man: John Cunningham Lilly. As a dolphin researcher I was of course aware of Lilly's monumental influence on the field of dolphin science—he is, as I am sure most dolphin scientists will agree, the father of the study of dolphin intelligence. His ideas and writings almost single-handedly propelled the dolphin to the forefront of the emerging field of animal cognition in the 1960s. But I was still astonished that so many of the concepts I was certain were attributable to later scientists (and non-scientists), like the idea that dolphins communicate by transmitting holographic sound images, were in fact first proposed by Lilly.

I want to make clear that this book is not a critique of John Lilly's ideas per se. The dolphin intelligence myth that Lilly originated has been evolving for decades as new scientific discoveries and fanciful speculation continue to churn up a heady soup of dolphin-inspired lore. And please be clear that when I use the term *myth*, I do not mean that this assembly of popular beliefs about dolphins is collectively false. Like all modern myths, and much like Lilly's own writings, when we look under the hood we find a complex web of truths and falsehoods, with uncertainties and speculation binding it all together like glue. Some of the concepts dealt with in this book are ones that Lilly himself might never have encountered. But Lilly's catalytic influence on dolphin science cannot be overlooked, which is why it's necessary to briefly recount just how he managed to transform what was initially regarded as an odd air-breathing fish at the turn of the 20th century into an animal whose intelligence is so sophisticated that it deserves the same constitutional protection as you or me.[12]

Lilly was a medical doctor and neurophysiologist with the National Institute of Mental Health. In his early research career he specialized in invasive cortical vivisection, which involved implanting electrodes into primate brains in order to monitor or stimulate the central nervous system.[13] Lilly's first encounter with a cetacean brain was in 1949, when he joined WHOI physiologist Pete Scholander on an excursion to examine the brain of a stranded pilot whale. Although the pilot whale's brain was too badly decomposed to conduct an investigation, Lilly's interest was piqued. On Scholander's advice, Lilly contacted Forrest G. Wood at Marine Studios in Florida, which he visited with a team of scientists in 1955 with the intent to map the dolphin cortex using the same techniques he had applied to primates. But his attempts at anesthetizing the dolphins failed—dolphins are conscious breathers, meaning that a fully anesthetized dolphin will rapidly succumb to asphyxiation as its brain loses communication with its diaphragm. Consequently, Lilly inadvertently euthanized five dolphins on this initial research excursion. But Lilly returned to Marine Studios in 1957 and again in 1958, having devised a method of inserting electrodes into the brain of a fully awake/conscious dolphin without killing it, and thus allowing his cortical mapping work to go forward.[14] He could then stimulate either the pleasure or the pain centers of the dolphin's brain as required by the experiment. This might sound barbaric to the modern reader—especially considering that these types of invasive procedure on marine mammals are no longer performed in the United States and a number of other countries. But keep in mind that at present, vivisection is routinely performed on many other animal species in US laboratories—from fruit flies to primates—so Lilly's techniques are not quite as outdated as they initially appear.

It was during one of these vivisection experiments that Lilly had his Eureka moment, which would result in a fundamental shift to how he—and the rest of the world—would relate to dolphin-kind. After reviewing slowed-down recordings of vocalizations produced by one of his dolphin subjects that had been making an awful lot of noise while having its brain stimulated just before it died, Lilly became convinced that the animal was attempting to communicate with the human experimenters by imitating the sounds of their speech. He also noted that when an injured dolphin was reunited with its tank mates, it "called" to them, and promptly received help/care—something Lilly argued was evidence of an intraspecies language. Lilly also observed that dolphins, unlike primates, did not become violent when having electrodes inserted into their brains—something he attributed to a sophisticated ability to control their emotions. The conclusion Lilly drew from these observations was that lurking inside the misleadingly piscine dolphin body was an undiscovered intelligence and capacity for language that could rival humankind.

Lilly presented his ideas on these subjects in short order at two scientific conferences in San Francisco in 1958, which drew the interest of Earl Ubell, the Science Editor at the *New York Herald Tribune*. Ubell organized a news conference for Lilly, which resulted in the dissemination of Lilly's novel ideas (which were eventually published in the *American Journal of Psychiatry*)[15] across the globe, and forever changed the nature of the relationship humanity had with the aquatic environment. The US Marine Mammal Protection Act of 1972—a landmark piece of legislation that banned "the act of hunting, killing, capture, and/or harassment of any marine mammal"—was due in no small part to Lilly's writings on the dolphin mind, and the effect his ideas had on public opinion. Here is an excerpt from Ubell's original article published soon after the news conferences of 1958, which likely constitutes the very first popular report of dolphin über-intelligence and the existence of dolphin language:

> Dr. Lilly believes they [dolphins] may be the only other creature besides man to be able to transmit complicated ideas by a kind of speech. And indeed they may be the only creature capable of learning true human speech. Next to man, it has the most complicated nervous system in the entire animal kingdom.[16]

After the flurry of media attention following his presentations, Lilly received grants to study dolphin behavior from the National Science Foundation, the Office of Naval Research, and NASA. The potential implications of Lilly's work as it pertained to controlling dolphin (and human) behavior via implanted electrodes caught the attention of the FBI, the CIA, the Defense Department, and J. Edgar Hoover, with Lilly being involved in classified security meetings to learn how, at the height of the Manchurian Candidate scare in the late 1950s, his research could apply to brainwashing and mind control. Although many scientists had initially criticized Lilly for going too far with his speculations concerning dolphin language and intelligence, the simple fact that the US government was interested in his work, which would eventually lead (as Lilly had correctly predicted) to the use of dolphins for mine hunting and other war-time applications under the US Navy's Marine Mammal Program, lent him considerable legitimacy in the eyes of the public. Lilly's scientific prominence was confirmed three years later when he attended a conference together with Carl Sagan, Frank Drake, and a dozen other leading scientists and deep thinkers at the National Radio Astronomy Observatory in Green Bank, West Virginia. This history-making conference established the field of SETI (Search for ExtraTerrestrial Intelligence) and the Drake Equation. The attendees became known as the Order of the Dolphin—presumably in deference to Lilly's influential ideas of communicating with other species—be it dolphins or aliens.[17]

By 1960, Lilly had set up a dolphin research facility (the Communication Research Institute) on St. Thomas, in the US Virgin Islands, and promised the world (and his financial backers) that he was on the verge of an interspecies communication breakthrough. But this breakthrough never happened, and Lilly's subsequent peer-reviewed publications were both sparse and underwhelming. By 1967, his funders had lost faith in both his ideas and his methods, and his financial support dried up. Lilly's critics from the scientific community—who had once described his highly influential 1961 publication *Man and Dolphin*[18] as an example on "how not to do scientific research,"[19] and suggested that the book did not present "a 'single observation or interpretation' that could withstand scrutiny"[20]—were somewhat vindicated by the ultimate fate of Lilly's research trajectory.

In his attempts to understand the dolphin mind at the Communication Research Institute, Lilly's experimental techniques became increasingly unconventional. He spent hours floating in an isolation tank under the influence of LSD, and even injected his dolphin subjects with LSD to see what would happen—a ludicrous notion to the modern reader, but perhaps only slightly unconventional in the zeitgeist of 1965. In his infamous experiments teaching Peter the dolphin to speak English, Peter was manually brought to sexual climax in order to satisfy his "sexual needs" and make him more cooperative[21]—something that is considered well outside the norm in animal behavior research, even for the 1960s. Lilly's exit from mainstream science saw him releasing three of his dolphins back into the open ocean (the other five having purportedly died of neglect), closing his laboratory on St. Thomas, and retreating to the West Coast, where he became one of the spiritual leaders (together with Timothy Leary) of the 1960s and 1970s counter-culture. Lilly was still involved in research projects involving dolphin–human communication (notably project JANUS in the late 1970s and early 1980s), but these projects did not resonate with the scientific community, nor result in any noteworthy scientific discoveries. He continued to publish hugely popular books weaving together his New Age ideology with dolphin science, providing the public with tantalizing yet wholly unsubstantiated statements like "these Cetacea with huge brains are more intelligent than any man or woman" and "the Cetacea are sensitive, compassionate, ethical, philosophical, and have ancient vocal histories that their young must learn."[22] Lilly is regarded by many as an unparalleled mythmaker (i.e., the man who planted "the 'mindbomb' of whale-dolphin intelligence"),[23] and is uniquely responsible for the emergence and intractability of many (if not most) of the ideas that will be scrutinized in this book.

Lilly's Lasting Legacy

It has been over half a century since dolphins were first declared language-using animal-savants by John Lilly. He once characterized himself as a maverick willing to stick his neck out[24] in defense of a position that the dolphin mind was superior to that of human beings. To those scientists uncomfortable with his habit of making grandiose and often unsubstantiated claims about dolphin intelligence, he once replied "only narrow-minded people criticize me."[25] Regardless of whether this attitude raises your hackles or fills your heart with admiration for Lilly's stand against orthodoxy, no one can deny that Lilly was an agitator par excellence; the man who pulled dolphins from their hidden aquatic realm and thrust them under the glaring lights of the world stage.

Famed cetologists Kenneth Norris and Karen Pryor, active during the heyday of Lilly-mania (the 1970s) and critical of his writings, cheekily referred to dolphins as floating hobbits.[26] Tolkien's mythical hobbits—much like Lilly's characterization of dolphins—are almost indistinguishable from human beings with regards to their complex behavior and intellectual prowess. Both hobbits and dolphins have become pop-culture icons, and stars of the silver screen. Fifty years after Lilly published *Man and Dolphin*, the super-intelligent-dolphin archetype that this book helped spawn is seemingly omnipresent in popular culture. Keanu Reeves starred alongside Jones, a cybernetic ex-Navy dolphin with cryptanalysis abilities surpassing that of humankind in the film *Johnny Mnemonic*. Darwin the dolphin from the television series *seaQuest: DSV* had his vocalizations translated into speech by the ship's computers, allowing him to conduct quick-witted conversations with his human shipmates. In a scientific review of how dolphins are portrayed in popular literature and media, two of the four most common dolphin themes aptly describe the fictional characters of Jones and Darwin, and seem to be derived directly from Lilly's own writings: 1) "Dolphin as peer to humans, of equal intelligence or at least capable of communicating with humans or helping humans," and 2) "Dolphin as superior to humans, associated with a higher power or intelligence."[27] It should come as no surprise that another major theme found by this study was "dolphin as representative of peace, unconditional love, or an idealized freedom in harmony with the natural order,"[28] which also has its origins in Lilly's early observations of dolphins' ability to remain calm and peaceful while having electrodes hammered into their brains.

It was not just popular culture that was influenced by Lilly's ideas, however. Lilly's contemporaries and subsequent generations of young scientists used his ideas as a springboard to test their own hypotheses as to the nature of dolphin intelligence. Consequently, other than the usual suspects involved in comparative psychology experiments (e.g., rats, mice, primates), perhaps no other non-human animal species has been subjected to as many different research paradigms intended to sniff out the nature of animal intelligence as have bottlenose dolphins. Whether we begrudge him or embrace him, no dolphin scientist can deny that the current body of literature on the subject of dolphin cognition is balanced firmly (if at times precariously) on the shoulders of John Lilly.

These days, most (although not all) dolphin scientists are comfortable with statements suggesting that dolphins are "highly intelligent" or have "sophisticated" or "complex" intelligence.[29] Sometimes modern researchers even pen statements similar in tone to early proclamations by Lilly that "it is probable that their [dolphins'] intelligence is comparable to ours,"[30] like the following:

> We can never fairly assess another species' use of the brain and mind. Such discussion tends to be left only for human-like relatives, based on our knowledge of one direction in evolution, that of land-based and human intelligence. Nonetheless, even in light of this concern, it does seem that dolphins should be recognized as members of the community of equals.[31]

Unlike Lilly's original claims however, these arguments are being offered in full acknowledgment of the difficulty of studying intelligence empirically, and in conjunction with decades' worth of scientific evidence suggesting that there might be something to the idea of dolphin intelligence after all. It is the modern science of dolphin cognition that has resulted in a number of recent media headlines that harken back to the early days of Lilly-mania: "New research suggests that dolphins are second only to humans in smarts," suggests Discovery News;[32] "Dolphins Are Animal Closest in Intelligence to Humans," states ABC News;[33] "Research reveals dolphins have extraordinary intellects and emotional IQs greater than ours,"[34] suggests the *Daily Mail*. These headlines are all from 2012, but are nigh indistinguishable from headlines that appeared half a century ago following the publication of *Man and Dolphin*.

The latest research into dolphin cognition and behavior has resulted in a new trend in public and scientific discourse that is threatening to do more than produce superlative headlines. The idea of extraordinary dolphin intelligence has formed the cornerstone of a number of legal initiatives around the globe keen on changing legislation in order to extend more rights and protection to dolphin species. Newly formed advocacy groups—some of them headed by leading marine mammal scientists—are campaigning for an end to captivity based on the idea that dolphins are too intelligent to be kept locked away. This includes The Kimmela Center for Animal Advocacy and the Nonhuman Rights Project,[35] which are exploring legal options to have bottlenose dolphins (and other animals) be given the legal status of personhood. On October 25, 2011, SeaWorld Orlando and SeaWorld San Diego were sued in the US federal court by the People for the Ethical Treatment of Animals (PETA),[36] with the backing of marine-mammal experts,[37] claiming that holding killer whales (the largest dolphin species) in entertainment facilities constitutes slavery in violation of Section One of the Thirteenth Amendment to the Constitution of the United States. The lawsuit[38] highlighted a number of facts pointing to the complex behavior of killer whales (i.e., orca), with a focus on intelligence, meant to justify their position as plaintiffs in the case, as in this example:

> The orca brain is highly developed in the areas related to emotional processing (such as feelings of empathy, guilt, embarrassment, and pain), social cognition (judgment, social knowledge, and consciousness of visceral feelings), theory of mind (self-awareness and self-recognition), and communication.[39]

The ethical and societal implications of sophisticated dolphin intelligence are continuingly being raised both in the courts and in the scientific literature. The question of whether dolphins are too smart for captivity was the focus of a series of impassioned commentaries appearing in *Science* in 2011.[40] Two special issues of the *International Journal of Comparative Psychology*[41] discussed the value and ethics of research on captive marine mammals, having taken its lead from a Humane Society of the United States' white paper that argued that it is unethical to keep an animal as intelligent and self-aware as a dolphin captive for entertainment purposes.[42] The 2012 American Association for the Advancement of Science conference held a symposium titled "Declaration of Rights for Cetaceans: Ethical and Policy Implications of Intelligence" at which scientists and philosophers argued that dolphins are intelligent enough to warrant granting them the same ethical and moral consideration as human beings.[43]

The question this book aims to answer is, how much of the current dolphin intelligence zeitgeist stems from good science, and how much is dubious fruit harvested from Lilly's original seed? Should these be considered frivolous lawsuits and spurious arguments that are making references to Lilly-esque pseudoscientific ideas, or is the current evidence for dolphin intelligence robust and indisputable, ready for its day in court?

The Intelligence Problem

Answering the question of whether or not dolphins are intelligent requires us to first address a rather fundamental problem: what is intelligence? And what do people mean when they proclaim dolphins as the second most intelligent species after *Homo sapiens*? I wanted to begin this particular discussion with a glib remark dismissing the idea that there is such a thing as the study of animal intelligence in the first place. As far as science is concerned, the idea that intelligence can be characterized as a single metric that allows for easy cross-species comparisons is a fallacy. Intelligence is an abstract and amorphous construct that, as Gould famously argued in *The Mismeasure of Man*, has been reified in defiance of both evidence and reason.[44] Gould characterized this form of biological determinism as leading us down an unscientific path where we attempt to distill intelligence down to a single value, and subsequently assign the worth of a human (or animal) based on this (arbitrary) value. The comparative psychologist Ed Wasserman advised that "it may be far simpler to provide precise operational definitions of elements or aspects of intelligence than to give a good definition of the overarching concept."[45] Following this advice, I thought it would make more sense to reject the term outright, and instead insist that we focus on the scientific study of animal cognition.

Marc Hauser offered such an approach in his book *Wild Minds: What Animals Really Think*, wherein he summarized the latest findings from the field of animal cognition, but purposefully avoided using the word *intelligence*, which he summarily dismissed as a concept having no value in the scientific study of the animal mind.[46] For comparative psychologists like Hauser and Wasserman, asking "what is intelligence?" is just like asking "what color is jealousy?'[47] It's not really a valid question, let alone a useful scientific approach to the study of the brain.

Cognition, on the other hand, is a hard science approach to what is often regarded as the soft science problem of intelligence. Cognition is the study of how the brain processes information, which we infer based on how an animal behaves, or even by directly measuring brain activity itself. This includes the study of perception, memory, categorization, learning, reasoning, communication, etc. The field of comparative cognition sprung up as a way of quantitatively comparing the many things going on in animal minds that might fall under the umbrella of intelligence. Unlike intelligence, cognitive processes can be precisely defined, and thus precisely measured scientifically. For example, one can study exactly which visual features of the face a sheep uses in order to recognize other sheep. It's then possible to draw meaningful comparisons between these cognitive processes in sheep and other species that also use facial stimuli to discriminate between different individuals (e.g., Capuchin monkeys).

The term *intelligence* is sometimes used to denote how complex an animal's behavior is judged to be. For a cognitive scientist, complexity can be a completely valid (i.e., objective) measure of either a static neurological system (e.g., an elephant's nervous system contains more complexity than the nervous system of a flatworm), or a behavior (e.g., humans' ability to send robots to explore the surface of Mars requires more behavioral complexity than a flatworm's autonomic reaction to a light source). Complexity within biological systems is merely expository, and is completely separate from any judgments as to which system or behavior is better. Evolution, the great arbiter of better-ness, is only concerned with fitness for a particular environment, and if having a complex brain that can create Mars rovers has allowed humans to thrive in our particular ecological niche, then so be it. But simple nervous systems churning out simple behavior, like those found in flatworms, have also led to numerous species that are thriving in their particular niches. In fact, the simplest organisms on the planet (i.e., the prokaryotes) are also the most abundant, and phylogenetically diverse. Having big brains or displaying complex behavior is by no means a one-way ticket to evolutionary success. When the term *intelligence* is used to describe complex behavior or biological systems, however, it almost always has a "better than" slant to it. It's impossible not to hear the *intelligence is good* subtext in the following statement from the BBC describing sheep facial recognition skills: "Researchers at the Babraham Institute in Cambridge, have shown that [sheep] have a remarkable memory system and are extremely good at recognising faces—which they suspect is a sure sign of intelligence."[48]

Science should not and cannot make claims about worth where brains or behavior is concerned, both human and animal alike. This is why cognitive scientists (like Hauser) are sometimes cautious when using the term *intelligent* to describe animal behavior. Consider the fact that ants, unlike sheep, use scent cues as opposed to faces to recognize individuals. From an objective standpoint, both scent recognition and facial recognition are likely a result of cognitive processes that are similarly complex. In a case where cognitive complexity is nearly identical, why then at first glance would most people, including me, intuitively consider face recognition a more intelligent behavior than scent recognition?

I am very sympathetic to the argument for tossing intelligence out the window as an undefinable, unquantifiable, and unjustifiable construct that brings with it a kind of worth/value judgment that has no place in science. But when it comes to the myth of dolphin intelligence, simply dismissing the idea as a scientific non-starter seems a bit like taking the easy way out. If an elementary school student were to ask me if dolphins are intelligent, and I were to launch into a rambling monologue about why intelligence is an untenable artificial construct, I would come off as the world's most despicable pedant. Regardless of whether or not intelligence is a useful scientific idea, I simply can't deny that even the most analytical scientist—just like any ten year old—has *a gut feeling* about what intelligence is. The challenge is explaining why we feel this way, and somehow funneling this answer into a working definition of intelligence that we can then use to answer the question of "are dolphins intelligent?"

Why Intelligence Matters

Intelligence is a concept with which we are all tacitly familiar. Humans have a natural inclination to classify and rank species on a scale of intelligence, even if we aren't really sure what it is we're comparing. We all seem to agree that intelligence is something arising in the brain that produces a specific kind of complex behavior, but it's hard to put our finger on exactly what this means. This is not unlike Supreme Court Justice Potter Stewart's characterization of what constitutes pornography; "I know it when I see it."[49] Our gut tells us that sheep's ability for facial recognition is intelligent, but ant scent recognition is not. A cognitive scientist might find this ridiculous, but they'd be in the tiny minority. A study found that Japanese and American university students consistently ranked species as to their perceived intelligence more or less in line with the putative "phylogenetic scale," starting with simple, single-celled creatures like amoebas, and progressing toward large-brained mammals like the great apes.[50] There seems something intuitively correct and even culturally universal about making the statement that mammals are, in general, more intelligent than insects, and we appear to do so without compunction. The idea of the *scala natura* is more than just an Aristotelian anachronism or defunct medieval Christian worldview; it appears to be the default approach humans have to classifying animals, their behavior, their intelligence, and, unfortunately, their worth.

For me (and most of my fellow humans), a fly ceaselessly bumping its head into a window pane is an example of unintelligent behavior, whereas a chimpanzee fishing for termites with a stick is an example of intelligent behavior. But how do we characterize this difference? Is this seemingly universal yardstick we're using to pass judgments on fly intelligence something that can be quantified? What specifically makes chimpanzee tool use intelligent behavior?

The most common ways we measure human intelligence, like psychometric tests for the illusive *g*-factor (e.g., Stanford–Binet), are generally useless when it comes to studying animal intelligence. Animals can't sit SAT exams, or (usually) tell you which number comes next in a series, and only specific aspects of tests that are independent of language or number knowledge (e.g., Raven's Progressive Matrices) can be given to animals. Despite the objections raised against trying to reduce intelligence to a single concept that applies to either humans or animals, plenty of scholars have offered operational definitions of intelligence that are intended to be universally applicable.

> The ability to attain goals in the face of obstacles by means of decisions based on rational (truth-obeying) rules.[51]
>
> Reason or intelligence is the faculty which is concerned in the intentional adaptation of means to ends.[52]
>
> Intelligence releases behavior from the direct control of the immediate stimulus and its history of reinforcement and permits flexible and adaptive solutions to novel problems and unfamiliar situations.[53]
>
> The ability to modify or create behaviour adaptively in the face of new evidence or changes in world conditions.[54]
>
> Intelligence may be defined and measured by the speed and success of how animals, including humans, solve problems to survive in their natural and social environments.[55]

These and related definitions of intelligence often circle around a number of themes: problem solving, reasoning, behavioral flexibility, ingenuity, etc. They often include the ability to think *about* information (as opposed to just *with* information), to manipulate it in the mind's eye, and to apply it to future behavior. This is the idea that intelligence is a kind of general purpose problem-solving mechanism in the brain. These definitions do a fairly good job of characterizing why a fly's head-bumping behavior seems so dim-witted.

But none of these definitions are going to work for us as we explore the five themes of the dolphin intelligence myth, since none of them fully encompasses the many behaviors displayed by dolphins that are suggested to be signs of dolphin intelligence—like those listed by PETA, or the ideas of dolphinese or dolphins' unusually complex social behavior. It's not really correct to describe having complex emotions like empathy, the capacity for self-awareness, or the ability to culturally transmit learned behaviors as falling in the category of *problem solving skills based on rational decision making*, and yet all of these attributes are regularly cited as signs of dolphin intelligence. To answer the question of how intelligent dolphins are, we are going to need a definition that neither replaces intelligence with a term like cognition (which is more scientific, but invalidates the question), nor uses one of the common definitions of intelligence with a narrow emphasis on problem solving. In my opinion, there is just one definition of intelligence that encompasses every single example of dolphin intelligence that I've seen in both the scientific and popular literature that includes large brain size, cortical complexity, tool use, flexible behavior, creativity, complex play behavior, problem solving, complex social structures, self-awareness, consciousness, complex communication, etc. This definition of animal intelligence also perfectly captures the essence of the universal yardstick by which we judge flies to be stupid and chimpanzees to be smart.

> Intelligence is a measure of how closely a thing's behavior resembles the behavior of an adult human.

When we ask how intelligent something is, we are really asking one thing: how human-like is this animal's (or robot's, or alien's) behavior? The reason this is an intuitive approach to the question of animal intelligence is due to what John S. Kennedy called "compulsive anthropomorphism"[56]—an almost irresistible tendency to attribute human-like thoughts and intentions to non-human animals. It is well known that the human mind attributes intentions, beliefs, and desires to a vast array of phenomena, including other humans, animals, inanimate objects, and natural processes like the wind, and sunsets.[57] The philosopher Daniel Dennett characterized this approach to understanding how objects in the world might behave as the *intentional stance*.[58] It is, arguably, one of the cognitive characteristics that allowed humans to succeed as a species in such a variety of ecological niches. It has tremendous predictive value, even if the things whose behavior we are trying to predict do not in fact have intentions, beliefs, or desires. A knock-on effect of this is that we intuitively judge the behavior we observe in the world around us as to how closely it resembles that of a fellow agent with a mind like ours; that is, another human. A chimpanzee behaves in ways more similar to what we'd expect to see in an animal that has human-like mental life. But a fly, with its head repeatedly crashing into a window pane, is displaying more than just a chronic inability to solve the problem of windows; it's acting decidedly un-human. This is the reason for the universal yardstick that places chimpanzees farther along the intelligence scale than flies—it's an anthropocentric approach to the world, and unabashed anthropomorphism. But because of our innate need for creating this type of hierarchy, the question of "how intelligent are dolphins?" is not a meaningless question that I should be replacing with a discussion of cognition. It is a legitimate concern for those of us living under the burden of the human condition.

This definition might explain why we have such a hard time classifying ant scent recognition as intelligent behavior. Although it's clearly a complex behavior, it's not particularly human-like, and thus does not bring the word *intelligence* to mind, or make us want to bestow the legal status of personhood on ants. Contrast this with the recent discovery that wasps can recognize each other's faces in a similar way to that of humans.[59] This discovery made front page headlines in December of 2011. It's the idea of an insect doing something "similar to humans" that causes us to sit up and take notice. It's the same argument as to why sheep's ability for facial recognition is worth noting: "If they can do that with faces, the implication is that they have to have reasonable intelligence. They are showing similar abilities in many ways to humans" suggested the scientists involved.[60] It's statements like these that pull back the curtain to reveal that intelligence truly is nothing more than a litmus test of human-ness.

I am not going to discuss whether or not anthropomorphism is good or evil as it relates to science, and whether or how much value it has as a tool for understanding the minds of non-human animals.

There are perfectly valid reasons why it should be avoided at all costs,[61,62] or embraced (in some measure) as a valuable tool in the ethologist's or comparative psychologist's toolkit.[63] Instead, I will throw my hands up and admit that when someone asks me about dolphin intelligence, the only answer I could possibly give that would make any sense is to discuss how human-like their behavior is. My personal opinion is that this is not a fruitful method of understanding or describing the cognitive abilities of other species, as it can bring with it unjustifiable teleological baggage, the scientifically bogus idea that some species are "more highly evolved" than others, and an inherent value-judgment. As Stephen Budiansky argued, there's no scientific justification for placing the "navigational abilities of pigeons or the web weaving of spiders or the nest building of bowerbirds or the food caching of nutcrackers" on a linear scale of intelligence where "humans equal 100."[64] But, as we have seen, this is the default method humans have for understanding the minds of other animals, and arguably the most common way humans judge the value of other species, so it cannot simply be brushed aside. And, more importantly, despite valid objections from animal welfare and animal rights advocates that ranking species by intelligence is impossible, it is dolphins' rank as more human-like than other species that is being proffered as an argument by advocates for "special attention," moral standing, and/or increased rights.[65]

Characterizing Dolphin Intelligence

So just how human-like is the behavior of dolphins, and is it more human-like than that of other species? An answer to this question will tell us whether or not it's correct to suggest that dolphins are the second most intelligent species after *Homo sapiens*. I will be looking at the most eye-catching examples of dolphin intelligence that have been cited in popular culture and the media to determine the extent to which these are based on solid scientific evidence. I will then compare what is known about each neuroanatomical, cognitive, and behavioral feature suggestive of dolphin intelligence to determine the extent to which it resembles its human counterpart, and whether or not it's unique to dolphins (and humans) and/or if it has been observed in other species.

Note that by comparing aspects of dolphin intelligence to that of other species, I do not intend to dismiss dolphins' "accomplishments" by carting out an ostensibly "unintelligent" species that does something similar to, or even better than dolphins. This was the abrasive approach that Paul Manger famously used to characterize dolphins as "dumber than a goldfish" following the publication of his controversial 2006 article on dolphin brain structure:

> You put an animal in a box, even a lab rat or gerbil, and the first thing it wants to do is climb out of it. If you don't put a lid on top of the bowl, a goldfish will eventually jump out to enlarge the environment it is living in. But a dolphin will never do that.[66]

In a *New York Times* op-ed piece in response to Manger's comments, the renowned ethologist Frans de Waal characterized this facetious approach as the false argument that "if a rat or pigeon can do it, it can't be that special."[67] De Waal is correct that this strategy for downplaying the behavior of dolphins by citing an example where another species (one putatively considered "dumb") does something similar or better than dolphins doesn't suddenly invalidate the body of research showing that dolphins display cognitive complexity in some domains. But if the argument is that dolphins are intelligent (i.e., similar to humans), and we discover that whatever behavior they are displaying that we once thought of as unique to humans (or unique to dolphins and humans), has also been observed in other species, then we could re-evaluate the trait in question as to if it should remain in our pantheon of *intelligent behavior*. For example: tool use, which had once been considered a trait unique to humans, then unique to primates, and then unique to mammals, has been found in species ranging from birds to cephalopods to insects. Consequently, tool use could arguably be struck off the list of uncannily intelligent human behavior, as it appears to be widespread in the animal kingdom. De Waal is entirely correct that this approach is bad science (and I am in complete agreement), but then again, comparative intelligence (unlike comparative cognition) isn't really science in the first place.

De Waal makes another important point concerning the trouble with comparing animal behavior: there can be vital differences in underlying causes of observed behavior that otherwise appears similar. A dolphin spontaneously inspecting itself in a mirror is clearly more suggestive of complex cognitive processes than a pigeon trained to use a mirror to peck at a spot on its chest. Nonetheless, as we shall see for dolphins, nearly every trait that we consider intelligent in dolphins has been discovered to a lesser or greater extent in other species, and, more often than not, we simply do not know how comparable the cognitive processes that underpin these behaviors really are. Octopus tool use is, by all accounts, equally as (or perhaps more) sophisticated than dolphin tool use. Chicken referential vocalizations are equally (or perhaps more) functionally referential than dolphin signature whistles. These findings must mean something about our understanding of animal intelligence, and cannot be dismissed as irrelevant because we assume dolphins are more intelligent than these other species and are thus relying on more complex cognition to achieve similar behavioral outcomes. Consequently, to fully review the evidence for dolphin intelligence, we need to put all of this in context by highlighting the ways in which dolphin behavior is (or is not) unique in the animal kingdom.

In order to get the facts out on the table, I will be compiling and analyzing evidence from a variety of sources. I aim to do this as impartially as possible, so that adversaries from both sides of any given controversy (e.g., pro- and anti-captivity) can hone their arguments by referencing facts that are universally accepted as being factual. This process will be guided by my sincere desire to let objectivity take center stage. I have friends and colleagues with strong and often opposing views on a number of contentious social issues, and if I can get each one to read this book and declare that "everything you've said about this topic is fairly presented," then I'll feel like I've done my job. I will not be playing the role of an advocate. My own personal feelings on any given controversial topic will, during the deconstructing process, take a back seat to my quest to uncover the truth.

Dolphin science, like all science, does not deal in certitudes, rarely producing the kind of concrete results that make for reliable front page headlines. Unfortunately, the public is perceived to be uncomfortable with the ambiguity stemming from research involving complex topics like the nature of the human and animal mind. Responsible scientists working in these fields engage in qualifying, hedging, and caveat-laden statement making when presenting the results of their findings to the public or speaking to the press, which can be frustrating for those looking for clear answers or easily digestible sound bites. Are dolphins intelligent? "Yes!" scream the newspaper headlines and advocacy groups. "It depends on how you define intelligence," drone the killjoy scientists. Instead of viewing this attitude as something problematic, I hope to inspire you to gleefully reject oversimplistic accounts of dolphin behavior and embrace the kind of ambiguity that drives dolphin scientists to engage in science in the first place. By providing as dispassionate and impartial an analysis of the current state of dolphin science as I possibly can, I hope to crack open the gaudy facade that shrouds our dolphin friends, and reveal a living, breathing, but deeply misunderstood mammal.

Lest you think I am on a campaign to knock dolphins down a peg in this intelligence discussion, let me put your mind at ease. After all the evidence is collated and assessed, it will turn out that dolphins (particularly bottlenose dolphins) are in fact much more similar to humans and other primates than one might expect of an aquatic mammal that evolved from a mammalian lineage that split from primates eons ago; our "cognitive cousins" as Lou Herman once called them.[68] However, unlike most popular science books on dolphins, I'm not simply going to cherry-pick all of the smart things dolphins do in order to reinforce the belief in dolphin super-intelligence. I will take the century-old words of famed psychologist Edward L. Thorndike to heart in noting that "most of the books do not give us a psychology, but rather a eulogy of animals. They have all been about animal intelligence, never about animal stupidity."[69] Thorndike was pleading for more rigor, objectivity, and fairness in the assessment of animal behavior, which is what I hope to provide in this book. Not by suggesting that dolphins are stupid, mind you, but by being honest about what the science is, or is not, telling us about dolphin intelligence. Let's begin by tackling the organ that gives rise to it: the brain.

2 What Big Brains You Have

Brains relate to cognition and mind, but no one is sure how.[1]

Steven Wise

Brain Size

The initial discovery that ignited Lilly's passion for the idea of dolphin intelligence was the simple observation that dolphins have large brains.[2] He argued that it is large brain size that leads to the ability to produce and comprehend language, culture, or any of the other intellectual feats made famous by *Homo sapiens*. By his reckoning, only elephants and a handful of cetacean species had brains large enough to harbor an intelligence that could rival that of humankind (chimpanzees were scratched off his list due to their small brain size). His decision to focus his research on bottlenose dolphins (brain weight of 1,824 grams),[3] which have larger brains than humans (1,500 grams)[4], but not African elephants (4,783 grams)[5] or sperm whales (8,028 grams),[6] was based on some rather odd reasoning, however.[7] Despite having brains three times larger than humans, an elephant is, he argued, unable to make sounds that resemble human speech, so there was no point in testing them for either language capacity or intelligence. Lilly concluded that sperm whales were the smartest animals on the planet based solely on the size of their brain. But he seemed happy to let this conclusion stand untested and unchallenged via experiment or observation because sperm whales are enormous, and thus might accidentally injure a human researcher.

Fast forward a few decades and we find that absolute brain size/weight had fallen out of fashion as a guideline suggesting the presence of intelligence in dolphins. As it turns out, brains, just like any other organ, tend to expand in proportion to the size of the animal, scaling allometrically with body size in a power relationship with a slope of approximately 0.75. Simply put: bigger bodies have bigger brains. There's certainly no evidence to suggest that, on the whole, increasing the size of the brain correlates with increased cognitive function. If there were, we'd expect larger animals to always be more intelligent than smaller ones. If brain size is the key to intelligence then cows (brain weight 423 grams) should exhibit complex cognition superior to gorillas (brain weight 406 grams).[8] But this is clearly not the case.

Of course sometimes, as in the case of humans, a large brain appears to go hand in hand with cognitive complexity. Humans have the largest brains of all primates, and we are (by our own reckoning) the most intelligent of primates. Absolute brain size can in fact be used as a predictor of some species of primates' performance on cognitive tasks, diet and foraging behavior in birds and primates, and social complexity in primates and ungulates. And, there is a modest correlation between absolute brain size and measures of intelligence in humans.[9] But large brains are not correlated with song complexity, food-hoarding skill, or cooperative breeding strategies and social complexity in birds. The occasional correlations between brain size and complex behavior are spread somewhat haphazardly around the taxa, and tend to break down when comparing traits between orders;[10] thus *universal* relationships between brain size and specific measures of intelligence are non-existent. So ambiguous is this line of research that some scientists have concluded that continuing to look for links between brain size and brain function is a colossal waste of time.[11] The underlying assumption in all of this, namely that increasingly complex cognition requires increasingly large brains, is by no means an established fact. Most entomologists would be happy to point out that insects produce a variety of complex behavior ranging from sophisticated social structure, numerosity, the ability to categorize objects as "same" or "different," recognition of other individuals, social learning, complex communication, etc., availing of the tiniest brains on the planet.[12]

Absolute brain size is not the only way of looking at the size issue, however. The funny thing about primate (including human) and dolphin brains is that, while they might not be the biggest brains of all, they deviate rather significantly from the allometric body-scaling rule. In other words, primate and dolphin brains are far larger than one normally sees in an animal of similar size—an observation that Aristotle made about the human brain some 2,400 years ago.[13] A number of metrics of brain size have cropped up that attempt to describe this anomalous observation, which, if humans are used as the baseline, might be a better predictor of intelligence than absolute brain size. This includes the calculation of brain weight as percentage of body weight, brain weight in relation to brain stem weight/size, and encephalization quotient (EQ), which is anthropologist Harry Jerison's measure of observed brain size relative to expected brain size for a given species.[14] Any of these measures have, at one time, been supplied as anatomical evidence for dolphin intelligence, but it is EQ that is most commonly cited these days. "If we use relative brain size as a metric of 'intelligence' then one would have to conclude that dolphins are second in intelligence to modern humans,"[15] argued Lori Marino, one of the leading authorities on dolphin neuroanatomy. Marino is referring to the discovery that many dolphin species have an EQ rating of between 1.55 and 4.56 (i.e., brains up to 4.5 times larger than the average animal of a similar size).[16] With modern humans having an EQ of 7, and primates having EQs under 3, this makes a handful of dolphin species (e.g., bottlenose dolphin, tucuxi, Pacific white-sided dolphins) second only to humans in terms of possessing a "relatively" large brain.[17]

But is it possible to use relative brain size as a metric for intelligence as Marino suggests? Is it correct, as the philosopher and dolphin rights advocate Thomas White suggests, that EQ "presumably correlates with the degree of behavioral complexity and cognitive capacity (what some might consider 'intelligence')"?[18] The quick answer is: the jury's still out. The science in support of this assumption is equivocal at best. For EQ to be a reliable indicator of cognitive capacity, it must have predictive value insofar as an EQ value should correspond closely with cognitive abilities in a robust, reliable, and reproducible way across the animal kingdom.

A handful of studies have indeed found correlations. Larger EQs are correlated with larger behavioral repertoire sizes (i.e., number of unique behaviors an animal produces) for a number of mammalian species.[19] A large EQ is also found in species that are better able to cope with and thrive in novel environmental situations.[20] Primates that feed on fruit as opposed to foliage also have higher EQs—ripe fruit being something that takes more brain power to locate than leaves.[21] Birds like parrots and corvids—which have larger EQs than other bird species—are generally considered the most cognitively complex. But for a meta-analysis of 44 studies of the relationship between EQ and the performance of 24 genera of non-human primates on nine different cognitive tests, it was found that EQ was not correlated with global cognitive ability.[22] A study looking at various primates' performance on measures of intelligence (including tool use, imitative ability) found that EQ does not correlate with or predict intelligence in primates.[23] For toothed cetaceans (including dolphins), larger EQs were correlated with larger pod size,[24] although whether or not the observed pod sizes used in the analysis were an appropriate indicator of the number of individual social relationships (and thus the increased cognitive power needed to maintain these relationships) is a matter of debate.[25] Strangely, the EQ for killer whales (2.57), which live in complex social groups (which is likely underpinned by complex cognition), is smaller than the EQ for harbor porpoises (3.15), which are generally considered to be a fairly un-social species.[26] There is, as of yet, no single cognitive trait that correlates strongly with EQ for cetacean species, let alone all the mammalian, non-mammalian vertebrates, or invertebrate species.

These conflicting findings suggest that EQ, like absolute brains size, is a fickle metric with little universal predictive value when it comes to estimating which species are going to perform well on any given cognitive test. Who could have guessed that African gray parrots like Irene Pepperberg's Alex, with a brain weighing just 9.18 grams (less than 1 percent of that of a human brain),[27] would turn out to be comparable to apes and bottlenose dolphins on tests involving symbolic communication comprehension?[28] Or that Asian elephants and rhesus monkeys (*Macaca mulatta*), both with an EQ of 2.1,[29,30] would pass the mirror self-recognition test[31,32] (although the rhesus evidence is controversial),[33] whereas stump-tailed macaques (*Macaca arctoides*), also with an EQ of 2.1,[34] would not?[35] Note also that capuchin monkeys, with EQ ratings ranging from 2.54 to 4.79,[36] which surpasses that of chimpanzees (2.2–2.5), do not perform anywhere near as well as chimpanzees on tests of symbol use, mirror-self recognition, etc.[37]

A recent study found that both primates and cetacean are unique in the mammalian order in that they yield the largest variation in EQ for species within that order.[38] A new hypothesis to explain this variance suggests that natural selection might have had a minimal role in the evolution of brain sizes for primates and cetaceans. Instead, neutral drift (i.e., changes to the genome/phenotype that have no influence on fitness) might be responsible for the over- and undersized brains for primates and cetaceans since both orders consist of species with small populations, which are known to be particularly susceptible to neutral/random drift.

The calculation of EQ itself has come under criticism, with one study showing that, depending on which species one uses in the groupings for EQ calculation, humans in fact have a normal brain size for primates of our body size, and it's our cousins the gorillas and orangutans that have strangely large bodies to house their otherwise normal sized brains.[39] EQ ratings can change depending on the data sources and different methods used for calculation, which means comparisons of ratings between studies might be providing spurious results. For example, human EQ ranges from 1.16 to 12.6 using the same data set but different calculation methods.[40]

As it stands, neither EQ nor absolute brain size have established themselves as metrics that can be confidently proclaimed as reliable predictors of animals' performance on cognitive tests associated with intelligence. Numerous studies on the relationship between brain size and cognition in animals have been conducted in the years since Lilly's first musing on the subject, but, while this seems a possibly fruitful line of inquiry, there is of yet no external validation of the relationship between the two.[41] The (relative) size of the dolphin brain is not, by itself, a variable that tells us anything concrete about the nature of dolphin intelligence.

Brain Structure

Aside from overall brain size, a number of structural properties of the dolphin brain have been suggested to be linked to intelligence and/or specific cognitive traits. An important lesson from the study of brain size is that, while brains do scale with the size of the body, they do not all scale up along the same lines. Different species with similarly sized brains might have very different looking brains from a neuroarchitectural standpoint, and it is the variation in these structures that many scientists argue make all the difference where cognition is concerned.

The cortex

The size and structure of the cerebral cortex (i.e., the outermost layer of the neural tissue covering the mammalian brain) is the natural place to start when looking for a link between brain structure and complex behavior. For humans and other mammals, the cortex, and in particular the neo-cortex, is the part of the brain most deeply associated with memory, reasoning, creativity, decision making, sense of self, language (in humans), emotion, abstract thinking, etc. In the book *Smarter than Man?* published in 1972, a pair of Swedish scientists concluded that, because the dolphin cerebral cortex is larger than that of humans, "it is nearly certain that dolphins and whales have a capacity for rational thought superior to that of man."[42] In reality, while the dolphin cerebral cortex has a larger surface area than the human brain (3,745 cm^2 vs. 2,275 cm^2),[43] it is considerably less dense, with far less brain tissue filling up all that extra space.[44] Regardless of its relation to the size of the human cortex, the dolphin cortex is undeniably large for a non-primate, or any animal for that matter. But what can cortex size tell us about cognition?

Modern scholars have suggested that the considerable size of the dolphin cortex suggests that dolphins have the brain structure necessary for complex cognition.[45] But what evidence is there for a relationship between neural hardware and complex cognition? Studies have found that cortex size is positively correlated with differences in social deception,[46] social learning and tool use in primates,[47] as well as social group size.[48] Paradoxically however, while both social deception rates and social group size are correlated with cortex size, they are not correlated with each other.[49] Nonetheless, a rather large body of evidence now provides support for the suite of related hypotheses suggesting that a large cortex size is directly related to the increased cognitive demands associated with living in large (and complex) social groups, with an emphasis on navigating tricky pair-bonds. Dolphin species produce a stunningly complex array of social organization, from the matrilineal family groups of killer whales to the fission–fusion societies seen in bottlenose dolphins, with long-term same-sex pair bonds lasting decades. Thus at first glance, the large size of the dolphin cortex would appear to be explained best by the *social brain hypothesis*.[50,51] But the relationship between cortex size and social complexity does not appear to be a universal rule in the animal kingdom. It holds up quite well when looking at primate brains, but when the net is cast a bit wider to include all species falling under the mammalian order Carnivora, the relationship disappears.[52] And it's not at all applicable to birds, some species of which can live in complex social groups and exhibit social learning, but don't possess a cortex. Thus, a large cortex does not necessarily go hand in hand with complex social cognition for all animals. It's a very attractive idea to think that the large dolphin cortex is directly linked to dolphin sociality, and this seems the most likely explanation at present, but like all brain/behavior correlations that are meant to encompass all taxa, this relationship is by no means an established universal truth. It's possible the large dolphin cortex might have evolved for reasons other than helping them keep track of their complex social lives.

One of the original hypotheses as to why the dolphin cortex needed to be so large was that extra cognitive power (e.g., speed, precision) was required to process echolocation signals.[53] Indeed for echolocating cetaceans, the auditory nerve and other mid-brain structures involved in sound processing are rather enormous, and the part of the cortex dedicated to auditory processing is large (consisting of two projection fields). But where the cortex is concerned, it's not the case that its size can be explained by auditory processing alone. In fact, a rather large portion of the dolphin cortex (i.e., the "association cortex") is involved in processing information not stemming directly from the senses, but rather connects the auditory fields with other sensory and motor areas.[54] The function of the association cortex is still unknown; it might be related to audio processing, but it might not, so it's impossible to pin the blame on audition for the overall size of the dolphin cortex. Consider that for other species with specialized sensory abilities that have led to enlarged neocortical areas (e.g., the platypus with two-thirds of its cortex dedicated to processing sensory information from its bill),[55] it's not the case that the rest of the cortex (or brain) grew relatively larger as well, as it did for dolphins. Consider also that Microchiropteran bats, which echolocate like dolphins, also have brains and cortical areas highly specialized for processing auditory information. Yet these bats do not have disproportionately large cortices or overall brain size like dolphins,[56] with many of them having EQ ratings below 1 (i.e., smaller than expected for an animal of their size).[57] The oddly oversized dolphin cortex might well be related to the demands of living in a complex social environment and/or the demands of processing complex echolocation and other auditory information (e.g., communication signals). But at present, there's simply not enough evidence to say with certainty that either of these two popular hypotheses are the primary reasons for the oversized dolphin cortex. Cortex size—like brain size or EQ—is yet another brain metric that does not have universal predictive power where animal cognition is concerned.

Number of neurons

It is frustrating to think that having a (relatively) large brain or large cortex doesn't always correlate with the production of intelligent behavior in animals. It seems intuitive that a large brain means a larger number of neurons, and thus increased information processing capacity, which should then result in complex cognition. But recent research suggests that this simplistic assumption is, despite its appeal, wrong. While having more neurons likely does result in more information processing capacity, it is not the case that bigger brains contain more neurons. Scaling up the size of an animal's brain does not always mean that the number of neurons responsible for information processing in that brain scale along with it. In 2005, Suzana Herculano-Houzel and her research team at the Universidade Federal do Rio de Janeiro developed a novel technique for obtaining accurate neuronal counts in animal brains, which led to the discovery that different orders of animals (e.g., rodents, primates) evolved unique strategies for fitting larger brains into larger bodies.[58] As rodent lineages evolved to be larger in size, their brain size increased more rapidly than their neuron count, resulting in decreasing ratio of non-neuronal to neuronal structures.

Consequently, large rodents have relatively fewer neurons for their brains size than smaller rodents. Primates, on the other hand, seem to retain the relative density of neuronal to non-neuronal structures as their brains increased in size, resulting in more neurons being packed into larger brains for the bigger primates. This is why a capybara—the largest rodent species—has a brain that weighs a hefty 76 grams, with 1.6 billion neurons, whereas the capuchin monkey, with a lighter brain weight of 52 grams, has over twice the number of neurons at 3.6 billion. For these two species, simply looking at brain size would leave us scratching our heads as to why capybaras are not more behaviorally complex than capuchin monkeys, whereas neuronal count provides us with a potential clue to capuchin intelligence. Using what is known of how primate brains scale up, the human brain actually contains the expected number of neurons for a primate brain of our size (86 billion neurons),[59] which appears to be more neurons than any other species, and might well explain why *Homo sapiens* are as intelligent as we are without having to suggest that the human brain is some sort of supernatural anomaly. For a rodent to achieve a similar neuronal count, their brains would need to weigh 35 kg—almost five times larger than the largest brain that has ever existed (i.e., the sperm whale).

This line of research also might be the best explanation yet as to how dolphin cognition and dolphin brain structure fit into the overall pattern of animal cognition. According to Lilly's ideas about big dolphin brains, or their stellar EQ rating, dolphins should be more intelligent than or just as intelligent as humans. But the current consensus is that dolphin cognition more closely resembles that of chimpanzees than that of humans. If there is a relationship between neuronal count and cognition, then it would not be surprising to learn that dolphin brains contain a similar number of neurons to their cognitive cousins the chimpanzees. At present however, we don't yet know how many neurons a dolphin brain contains, although Suzana Herculano-Houzel's research team is working on this problem as I write these words. If delphinind brains scale in size similar to rodent brains, dolphins should have a neuronal count similar to the non-human great apes. But if they follow the primate brain-scaling pattern (i.e., densely packed neurons), their neuronal count could be astronomical. Best guesses at the moment suggest that dolphin neuronal count will be three-quarters that of humans (on par with chimpanzees),[60] since it is already well established that the dolphin cortex is considerably less densely packed with neurons than that of primates, and that, like rodent neurons, dolphin neurons are relatively large in size.[61] It has already been hypothesized that the dolphin brain grew large in order to obtain a larger neuronal count (and thus more cognitive complexity) by means of duplicating relatively thin cortical areas resulting in a large (but thin) cortex, as opposed to packing more neurons into the cortex as observed in primate brains.[62]

It is too early to say for sure how well absolute neuronal count matches with cognitive abilities across taxa, and what this means for dolphins, as this research is still in its infancy. Neuronal count might well be the best all-around predictor of animal intelligence as some researchers have suggested,[63] or it might reveal itself to be yet another unreliable metric (as others have predicted)[64] in the ongoing quest to understand how brain structure correlates with intelligent behavior. After all, it will not likely solve the persistent paradox of small-brained animals with small neuronal count (e.g., rats, ravens, honeybees, octopuses) displaying unexpectedly complex cognition.

Smart neurons

Simply looking at the absolute number of neurons in a brain for clues to intelligence is somewhat akin to looking at the number of soldiers each army has on a battlefield for clues to the potential victor. For both neurons and soldiers, it's likely not simply the number of boots on the ground that matters (although this is obviously important), but their speed, efficiency, and specialization as well as their ability to communicate with each other and organize effectively. A handful of neuronal structural properties have been singled out as potential signposts leading to intelligence. The number of synapses that neurons possess (similar in number for humans and dolphins)[65] might correlate more closely with complex cognition than neuronal number. Or perhaps it's the length of the neural pathways between different brain regions that is the key to intelligence—pathway length can affect the speed of signals and overall communication efficiency of the brain. Or maybe neural plasticity—the ability to adapt neuronal connections to environmental input—is the most important predictor of cognitive complexity. Or maybe the ratio of glial cells to neurons is important. Glial cells were traditionally thought of as the glue that holds together the neurons in the brain, although their role is far more important than that; they might even be involved in the transmission of information.[66] Dolphins likely have a relatively high number of glial cells in the cortex,[67,68] which some neuroanatomists have argued is an indication that dolphin brains are well equipped for complex cognition. Of course, others have argued that this indicates the exact opposite.[69]

Perhaps one of the most eye-catching discoveries where cell structure and dolphin brains are concerned was the revelation that dolphin brains contain von Economo neurons (VENs), previously known as spindle cells. VENs have been singled out as a neuroanatomical basis for complex cognition in dolphins. As the venerable Stephen Fry crooned in his narration of the 2011 BBC documentary *Ocean Giants*, "The latest research on whale and dolphins' brains has revealed something quite unexpected. Like us, they have spindle cells. These special brain cells were once thought to be unique to humans because of their link with language, self-awareness, and compassion."[70]

VENs are long, thin neurons with one dendrite at each end (as opposed to several branches) that act like "express trains" of the nervous system.[71] In humans, VENs are concentrated in the anterior cingulate, anterior insular, and frontopolar cortices, regions of the brain known to be involved in emotional and social awareness,[72,73] and possibly the seat of complex social emotions like embarrassment and empathy. It is hypothesized that VENs help to quickly communicate thoughts of self- and social awareness from these three cortical areas with the rest of the brain, including subcortical areas. It is hypothesized that these connections might make us more aware of—and in control of—our own visceral reactions, emotions, and vocalizations, and aid in decision making.[74]

Once considered a neuronal cell structure unique to humans and great apes (as Stephen Fry pointed out),[75] VENs were discovered relatively recently in the brains of elephants[76] and large cetaceans, including humpback whales, sperm whales, fin whales, and killer whales.[77] Because these species are large-brained mammals that possess fairly complex social structures, researchers hypothesized—and the media reported—that "smart, social animals such as whales and elephants might have the same specialized wiring for empathy and social intelligence as human beings."[78] When VENs were discovered in the brains of a number of dolphin species soon after,[79] VENs were treated as direct evidence by many of the neuroanatomical underpinnings of complex social emotions in animals: "Dolphins and whales are known to be capable of empathy because they have the same spindle cells in their brains as humans," reported the *Daily Mail*.[80]

The evidence of the functional role of the brain areas in which VENs are found in humans is fairly strong. When these brain regions are damaged due to Alzheimer's disease, the behavioral variant of frontotemporal dementia, autism, or schizophrenia, impairment in social cognition, empathy, self-awareness, and social judgment can be observed.[81] Studies confirm that these brain regions are responsible for similar behavior in primates.[82] Presumably, VENs located in these areas must be vital to the function of these cortical areas, and thus the production of complex social cognition. But is the presence of VENs direct anatomical evidence of a capacity for similar complex cognition in dolphins and other animals as some have argued?[83] This might be too much of a leap. To begin with, the organization of the dolphin cortex is so strikingly different to that of primates that it is impossible to state with certainty that the anterior cingulate, anterior insular, and frontopolar cortices in dolphins share a similar function to the same cortical regions found in the great apes. And even for humans, there are examples of patients with severe trauma to these brain regions who nonetheless retain capacities like self-awareness which, given our current understanding of the function of the regions, should be impossible.[84] Also, it's important not to overstate what is known about the exact function of VENs—science is still not really sure what purpose they serve in these brain regions. As the scientists who first discovered VENs in dolphins noted, there is no direct evidence of the functional role of VENs.[85] It's tempting to see them as the cells responsible for rapid communication between brain regions facilitating complex social cognition, but, as Frans de Waal cautions, "until someone establishes the exact function of those cells, it remains a story, basically."[86]

The story is complicated by the fact that VENs have recently been found in the brains of species that are, by all accounts, not particularly known for being socially or cognitively complex, which makes it even harder to argue that VENs are somehow responsible for complex social cognition. This includes Florida manatees, walruses, zebras, black rhinoceroses, domestic horses, and the pigmy hippopotamus.[87] VENs were found very recently in macaques,[88] which upsets the traditional view that VENs are unique to the great apes within the primate lineage, although it is noteworthy that one macaque species (*Macaca mulatta*) might have passed the mirror self-recognition test.

It's an important point, however, that VENs that have been discovered in the great apes, cetaceans, and elephants all seem concentrated in the anterior cingulate, anterior insular, and frontopolar cortices, whereas VENs are spread somewhat haphazardly, and sparsely, around the brains of species like manatees. This might provide an important clue as to why these animals in particular may be benefiting from the presence of VENs in their brain, and we can look to similarities in the domain of social cognition as a possible explanation. But not all of the species that possess these cortical areas together with VENs are necessarily all that similar in the area of social cognition to begin with. Consider that the fin whale, a species that is regularly lumped in with their distantly related cousins the bottlenose dolphins when talking about VENs' role in social cognition, does not readily fit the bill of a socially complex marine mammal. Unlike bottlenose dolphins, which lead complex social lives involving life-long pair bonds and are known to coordinate complex hunting strategies, there is little evidence that fin whales' social systems are equally as intricate. Thus a link between VENs, complex social cognition, and EQ has not been established for fin whales. And fin whales, like manatees, also suffer from having particularly low EQs (0.49[89] and 0.27[90] respectively), which also does not fit with the larger narrative about EQ and intelligence, making it difficult to tie a neat bow on the idea that VENs are part and parcel of the big brained, socially complex species paradigm that otherwise fits so well for bottlenose dolphins or chimpanzees. Consider also that, of all the animals discussed so far, the one with the largest distribution of VENs throughout the cortex is the pygmy hippopotamus. Looking at the bigger picture of which species possess VENs, how many they possess, and where they possess them, it's clear that we lack a coherent evolutionary explanation as to how, why, and when VENs evolved, let alone what their function might be.

It is entirely possible that VENs are types of neurons that, regardless of their potential use in social cognition, crop up spontaneously when brains grow beyond a certain size (like the large brains of great apes and cetaceans) as a means of facilitating communication between brain regions—an idea that has been proposed in many forms before.[91,92,93,94] Indeed, VENs appear to have evolved independently in a number of large-brained lineages. Of course, VENs are also found in the brains of animals with minimal cortical folding and a small overall size (e.g., manatees), so perhaps brain/cortex size is not the only factor driving their evolution. It's also worth considering that the function of VENs might be very narrow, facilitating speedy transfer of information between specific cortical areas involved in just a single function—like monitoring of visceral function—and not something as broad as "social cognition," which includes a host of cognitive abilities.

The all too familiar caveat for brains remains: we don't know exactly how these structures relate to cognition. While some suggest that the presence of VENs is support for the idea that dolphins are capable of empathy and higher-order thinking,[95] it's probably more accurate to state that "the presence of these neurons in cetaceans does not demonstrate, but is consistent with, complex cognitive abilities."[96] In other words, you might expect to see VENs in the brains of species with complex cognitive abilities, but finding VENs is not, on its own, evidence that an animal will be cognitively complex. In a similar vein, all birds need wings to fly, but this does not mean that when you find a bird with wings, it can fly (e.g., penguins, ostriches). On the flip side, consider that there is a lot of cognitive complexity—even in the domain of social complexity—displayed by species that probably lack VENs (or at least very many VENs), including ravens, magpies and other corvids, hyenas, wolves and other canids, parrots, baboons, etc.[97] Following from the wing–VEN analogy, this would be akin to finding a species that had no wings, but could nonetheless fly. Some of these species, like the corvids, do not even possess brain areas that are homologs or analogs to the anterior cingulate, anterior insular, and frontopolar cortices; not only do they not possess VENs, but they are producing cognitive complexity with completely different parts of their brains. It's also not clear just how different VENs are from other pyramidal neurons, and entirely possible that, in species where we do not see VENs that resemble human VENs, there are nonetheless morphologically related neurons carrying out an identical function.[98] So the presence or non-presence of VENs doesn't appear to be telling us anything concrete as far as social cognition in the animal kingdom goes. All we can really say at present is that the presence of VENs—like any other brain structures—are a mysterious signpost that science is yet unable to decipher.

In summary, studies show positive correlations between a number of measures of neuronal structure and cognition for both humans and non-human animals, yet none of them is likely to be applicable to all brains across all taxa. It could well be that the whole business of measuring and counting neurons, synapses, VENs, neural path lengths, plasticity, glial cell ratios, neuronal connectivity, etc., in search of the root cause of intelligent behavior is currently a red herring—providing us hints at possible relationships, but ultimately drowned out by the chaos that persists when attempting to pair cell structure with cognitive function. I personally believe that science will one day make sense of this chaos, and that this is in fact a vital research path. But if we are being truly honest with ourselves vis-à-vis the modern science of the brain, simply looking at the cytoarchitecture of the dolphin brain tells us frustratingly little as far as intelligence is concerned, however tantalizing the clues might be.

Smart structures

Moving up from cell structure to gross brain anatomy, we notice that the dolphin brain—when compared to primate brains—contains a number of differences in general morphology, and the presence and relative size of a number of structures involved in cognition. For humans and other primates, the frontal lobes—especially the prefrontal lobe—is considered the seat of complex cognition, where executive function, the sense of self, and important aspects of consciousness likely reside. Dolphin brains did not evolve an analogous frontal lobe, but instead saw expansion of the cortex in lateral areas, resulting in a brain that is rounded as opposed to elongated like that of primates. Quite possibly the cortical areas responsible for the kinds of cognition seen in the frontal lobes of primates have, for dolphins, simply been shoved to lateral areas due to the way the dolphin brain evolved to accommodate the nasal passage and telescoping of the skull.[99] But the lack of a distinct frontal lobe (the reduced dolphin equivalent being relabeled the orbital lobe or frontopolar cortex) has prompted some scientists to suggest that dolphins cannot possibly be intelligent or self-aware animals.[100] Of course, plenty of experimental evidence shows that dolphins produce complex behaviors (e.g., mirror self-recognition) that have been historically thought of as the product of a frontal lobe that they do not in fact possess. It's difficult to say what having a frontal lobe (or not having one) means as far as intelligence goes; ravens and parrots have a larger forebrain structure (somewhat equivalent to a frontal lobe) than other birds, which might explain why they seem to be the most cognitively complex birds. But the traditional view that the human frontal lobe is relatively large compared with other primates, and thus results in our intelligence, has been challenged by recent evidence.[101] In the end, the importance of having some sort of bulging front-brain structure in order to facilitate complex cognition is by no means a hard and fast rule. If anything, dolphins are evidence that it's probably not a rule at all.

Other anatomical oddities of the dolphin brain include a relatively small hippocampus, and a relatively large cerebellum. But what this tells us about dolphin cognition is, as I suspect you will have guessed by now, not at all clear. The hippocampus is integral to the processing of memory and learning for most species, particularly in the area of spatial navigation. It was famously discovered that London taxi drivers have enlarged hippocampal structures, with the increase in volume directly correlated with how long they had been driving a taxi.[102] A relatively large hippocampus has been strongly correlated with birds' abilities to store and retrieve food. But despite the relatively small size of (parts of) the dolphin hippocampus, they seem particularly skilled at tasks involving learning and spatial navigation, which leads to the conclusion that hippocampal size is not a significant predictor of these skills for dolphins, and that their brains must be organized to cope with these cognitive skills in a fundamentally different way to terrestrial mammals and birds.

The dolphin cerebellum is much larger in relation to other brain areas in dolphins than in primates. The cerebellum's function is generally thought of as controlling and coordinating movement, as well as monitoring sensory information. It might then be that dolphins need a larger cerebellum to help process acoustic information from their echolocation while tracking their prey. The cerebellum has also been implicated in tool use for primates, so an enlarged cerebellum in dolphins might be related to their ability to use tools.[103] Of course, octopuses have been observed using tools,[104] and they, technically, don't have a cerebellum. And, as it turns out, the clearest relationship between tool use and neuroanatomy in animals is the amount of cortical folding, and not cerebellum size.[105] So exactly what a large cerebellum means for dolphins is unclear.

The inescapable problem with comparative neuroanatomy is that it seems that robust links between the presence or size of one specific brain area and performance on specific cognitive tasks across taxa is all too rare. Similar brain regions in different species might serve vastly different functions. But even if function were conserved for analogous brain regions across species, consider that for complex behavioral tasks, numerous brain regions are involved, and the relative importance of one region is not immediately clear. For example, when macaques use tools as opposed to simply manipulate objects, a complex series of ten different brains areas are involved, including the basal ganglia, premotor cortex, cerebellum, etc.[106] This makes it very difficult to point to individual brain structures as being the driving force behind specific behaviors for a single animal, let alone allowing us to make sweeping generalizations about the function of specific regions for all animal brains.

Putting the brain in context

Just what are we to make of this confusing list of dolphin brain properties and their seemingly elusive relationship to cognition? In public discourse, the oversimplified arguments for a link between brain structure and intelligence in dolphins sometimes careen into unfalsifiability. Where dolphin brain structure resembles that of humans (e.g., high EQ, large cortex, presence of VENs), it is touted as direct anatomical evidence of their intelligence. Where brain structures differ in fundamental and bizarre ways (small neuronal density, lack of distinct frontal lobe, small hippocampal structures), it is touted as evidence of how intelligence can be produced by a distinctly un-human-like brain. In both cases dolphin intelligence is the a priori conclusion, and is unchallenged by whatever brain structure we find inside a dolphin's skull. Consequently, brain structure can at times be a footnote to behavioral evidence of intelligence, as opposed to something explanatory or predictive on its own. In a general sense, it seems correct that a number of metrics of the dolphin brain point to a level of complexity that is similar to that of primate brains, which is consistent with the idea that dolphins and primates both exhibit complex behavior stemming from complex brains, possibly a result of something akin to convergent evolution. But that is not the whole story, and when looking closely at the anatomical details, and broadening our view to include the brains and behavior of other species, universal links between specific brain metrics, neuroanatomical properties, and behavior seem to slip from our grasp.

The pitfalls of using brain size or structure as direct anatomical evidence of intelligence are brought into sharp focus by the now infamous story of Paul Manger, Professor of Anatomical Sciences at the University of the Witwatersrand. In 2006, Manger published an article outlining his ideas as to why the dolphin brain evolved to be so large.[107] He presented three main points: 1) cetaceans evolved large brains designed for thermogenesis (i.e., heat production) as evidenced by a decrease in ocean water temperature that coincided with the increase in cetacean brain size approximately 30 million years ago; 2) the neuroarchitecture of the cetacean brain appears to be designed for thermogenesis and not information processing; and 3) the research results claiming that dolphins are intelligent based on observation of their behavior are based on unproven assumptions and, in actuality, dolphins are not particularly intelligent, as intelligent behavior is the result of complex neural processing for which there is no evidence in the brains of cetaceans.

These ideas created an immediate controversy, and a heated response from the scientific community involved in dolphin brain and behavior research. A number of subsequent articles directly challenged Manger's ideas concerning the relationship between brain size and thermogenesis as it pertains to cetacean evolution.[108,109,110] But importantly for this discussion, Manger's claims about the relationship of neuroanatomical structure to cognitive complexity received much scrutiny. He offered a number of examples of why the structure of the dolphin brain suggests they cannot be intelligent animals, including:

1. low cell (neuronal) density
2. thin cortex
3. low number of cortical areas
4. small prefrontal cortex
5. high glia–neuron ratio
6. small hippocampus.

We've seen that for each of these points, the current state of the science suggests quite clearly that science has no real idea what these observations might mean as far as cognition goes, or even if these observations are necessarily correct. Each one of these points could just as easily be linked with dolphin intelligence as opposed to dolphin stupidity, which is why both Manger and his opponents are able to offer seemingly endless reasons why their position should be regarded as the most parsimonious. The strangest aspect to Manger's argument, in my opinion, was his insistence that the experimental and observational evidence of intelligent dolphin behavior should not be counted as evidence of complex cognition. I once interviewed Professor Manger for a podcast I produced on this subject, and asked him the following question: "In your opinion, is there any convincing behavioral evidence from experimental research with cetacean species (particularly odontocetes) that might suggest that they indeed have cognitive abilities on par with or exceeding that of primates?" Manger replied simply: "no."[111]

It was this position that most incensed the dolphin science community, with subsequent articles published in reply to Manger's ideas outlining example after example of seemingly undeniable parallels in experimental and observational results of dolphin and primate cognition.[112,113] Manger insisted however that if the brain structure for fast information processing was lacking in dolphins (although many argued it wasn't), then they simply could not be considered intelligent animals.[114] Lance Barrett-Lennard, a marine mammal expert, provided a pithy but deadly accurate summary of the mainstream scientific community's response to this particular point: "A dolphin could have a brain the size of a walnut and it wouldn't affect the observations that they live very complex and social lives."[115]

Which brings us to the take-home message for this discussion of dolphin brain anatomy. The reality is that science is currently almost entirely in the dark when it comes to a nuanced understanding of how brain structure relates to behavior and cognition in a global way, and it is at best premature and at worst disingenuous to suggest that we can say anything conclusive about dolphin intelligence by examining dolphin brains under a microscope. Even for something as well studied as the human brain, science is obliged to conclude that no anatomic feature(s) can adequately explain the nature of human cognition, consciousness, or the specifics of our intelligence. The diversity of brains in the animal kingdom contributes to a diversity of behavior, with links between the two being tenuous at best. Perhaps in a few decades (or perhaps centuries) science will be able to speak with a bit more authority on this topic, but as far as a discussion of dolphin intelligence goes at present, one can argue that we really shouldn't be focusing so much of our attention on the walnut. Lou Herman's edict, penned over three decades ago, still applies: "It is behavior, not structure, that measures the intellectual dimensions and range of the species."[116]

Cogito Ergo Delphinus Sum

A man with so large a brain must have something in it.[1]
Sherlock Holmes

Body and Mind Knowledge

The idea that dolphins are self-aware and/or conscious is often the leading argument as to why they deserve some form of moral standing, the metaphysical status of personhood, or possibly the legal status of personhood. In summarizing the research into the dolphin mind, the philosopher Thomas White suggested that "the science has shown that individuality, consciousness, self-awareness is no longer a unique human property,"[2] and that the intelligent dolphin brain "appears to support a consciousness" that lets dolphins "be aware of themselves and others."[3] This forms part of his argument as to why dolphins should be considered non-human persons.

For many cognitive scientists, however, the problem of self-awareness in animals remains one of the most contentious issues in the study of animal behavior. It involves a minefield of complex and poorly understood psychological concepts concerning the mental lives of dolphins as it pertains to their awareness of their bodies, actions, minds, thoughts, and emotions, as well as their awareness of the minds, thoughts, and emotions of other beings. Throw in the problem of what consciousness is, and you've got yourself one of the thorniest issues in the history of science and philosophy. If science has anything concrete to say about the dolphin mind with regard to these topics, it's that we've found some tantalizing clues suggesting that dolphins have some sort of awareness of their own minds (and possibly the minds of others), but we don't know for sure how similar this is to human awareness, nor how unique the dolphin mind might be when compared with other species on these points.

There's simply no consensus at present concerning how we should define different levels/kinds of self-awareness, or what the best methods are for determining the presence of self-awareness in either humans or non-human animals. Studying consciousness or awareness in animals is more or less a brand new field of empirical inquiry that has only recently shrugged off the criticisms by those who steadfastly refused to believe that studying the mental experiences of animals was either possible or necessary.[4] Consequently, the evidence we have for awareness in dolphins stems from just a decade or two of investigation, and avails of methods (e.g., the mirror self-recognition test) that are still being debated and validated. We are in the early stage of consciousness science where animals are concerned, which means that nothing is certain and there are no guarantees. Thus, the results of tests for body and mind knowledge in dolphins are currently inconclusive or open to interpretation, with virtually nothing lending itself to an unequivocal statement about what "the science has shown." With these caveats in mind, let us now see exactly what the science is telling us about dolphin self-awareness at this moment in time.

Body-awareness

Perhaps the most basic form of self-awareness is the knowledge that one has a body that is separate from other objects encountered in the world. This body awareness (sometimes called *embodiment*)[5] relies on a sensory–perceptual system that lets the brain know where the body (or parts of the body) are in space (e.g., proprioception, kinesthesia), in addition to the many afferent sensations (e.g., pressure, temperature). This sense of body-awareness can arise in the mind independent of the immediate sensory information from the body itself, and is thus an important first step on the path to having an emergent sense of self. This is evidenced by a number of strange syndromes in humans where a disconnect occurs between the mind's awareness of one's body plan or body schema, and the actual sensations the mind receives from the body. *Phantom limb syndrome* occurs when a sufferer has a limb amputated but retains the experience and sensation of the limb; *alien hand syndrome* occurs when a sufferer's hand feels like it is moving on its own accord (i.e., against their will) even though they retain both sensation of the hand and a feeling of ownership of the hand; and *body integrity identity disorder* occurs when sufferers feel that one of their limbs (which they can sense and move at will) is not actually part of their own body, and often attempt self-amputation to relieve their suffering.

Evidence from the laboratory suggests that dolphins possess a sense of body-awareness. At the Kewalo Basin Marine Mammal Laboratory run by Louis Herman, the bottlenose dolphin named Elele learned symbolic gestures that represented nine different body parts: rostrum, mouth, melon, dorsal fin, side, belly, pectoral fin, genitals, and tail.[6] When given instructions to show or shake different body parts, or use different body parts to carry out actions (e.g., touch the Frisbee with your dorsal fin), she was able to do so. These results demonstrate that Elele was able to form conceptual representations of her various body parts, including parts she was not able to see visually (e.g., her melon), which, given her ability to carry out instructions involving these representations, might suggest that Elele had some kind of conscious control of her body resulting in a kind of body-awareness not unlike that of a human.[7] The language trained great apes (e.g., Kanzi the bonobo)[8] also learned symbols referring to their body parts, but have not been tested to the same extent as Elele in using body parts to carry out actions.

Agency and imitation

Moving up from body-awareness we find the concept of agency, which in its most basic form is the sense that one can control one's movements. There is debate in the literature as to how proprioception, sensory information, and motor movement and feedback combine to create both a sense of agency and of awareness of body, and whether or not agency and body-awareness might be mediated by separate pathways in the brain. When things go wrong, as in the case of schizophrenia, one might retain a sense of body-awareness, including being conscious of one's movements and actions, but have lost a sense of agency insofar as one feels that "someone else" is responsible for controlling one's actions. When things go right, body-awareness and agency create the bedrock upon which an animal can make a distinction between self and other, and ultimately leads to more complex forms of self-awareness.

Perhaps the best means of testing for an animal's sense of agency involves studies of imitation. Imitation is defined as "accessing a mental representation of an experienced event to reproduce that event through one's own behavior."[9] Lou Herman's language-trained dolphins from Kewalo Basin were able to follow a command to repeat a behavior they were asked to perform,[10] as well as repeat a novel behavior that they had self-selected.[11] This evidence suggests that these dolphins were able to monitor, remember, and self-imitate their own actions—which most certainly requires some form of agency. Recent evidence shows that killer whales are equally as skilled at quickly learning to imitate both familiar and novel body movements of other killer whales.[12]

Dolphins' ability to imitate both actions and sounds occurs in a wide variety of contexts. Dolphins spontaneously mimic sounds in their environments, artificial and computer generated sounds, and each other's whistles.[13,14,15] Killer whales will both imitate and learn the vocalizations of their tank mates (in captivity), as well as other family groups with which they come in contact in the wild.[16] Killer whales have been observed imitating sea lion barks,[17] and bottlenose dolphins imitate humpback whale song.[18] Guyana and bottlenose dolphins will mimic each other's vocalizations when engaged in agonistic encounters.[19] Dolphins can imitate the body movements/behavior of both other dolphins and human beings, including when displayed on a video screen.[20,21]

Aside from imitation's ability to reveal a sense of agency, it might be implicated in a variety of socio-cognitive knowledge including establishing the self-other distinction and awareness of the minds of other agents. It is also vital to social learning, culture, and the ability to adapt to novel situations. There are a variety of levels of imitation, which range from mapping the body plan of another agent onto one's own body plan in order to repeat an action (i.e., kinesthetic imitation), to guessing the action that another agent intended to perform, which results in imitating the same goal if not the exact same action (i.e., true imitation).[22,23] All forms of imitation involve some form of mental representation, but it is not quite clear how rich these representations (including self-representation) might need to be. Kinesthetic imitation in human infants appears as early as six weeks, which suggests it is a product of low-level cognition in humans. While it might be impossible to say when dolphin imitation has moved from kinesthetic imitation to true imitation, and thus involve a more complex mental representation of the intentions of other agents, there is no doubt that dolphins are most certainly aware of and in control of their own actions in a way that is commensurate with a definition of agency.

Of course, dolphins are not the only animals displaying imitative behavior. The vocal imitative abilities of birds (e.g., parrots, lyre birds) are perhaps unsurpassed in the animal kingdom. Elephants are able to imitate the sounds of cars/trucks, and African elephants have been observed imitating the vocalizations of Asian elephants.[24] Goats might be availing of some form of vocal plasticity (if not learning and imitation) in that they can modify their vocalizations to more closely resemble the vocalizations of their social group.[25] Bats and seals/sea lions too display vocal imitation when developing their call repertoire.[26,27] And, of course, the great apes (unlike monkeys) are particularly skilled at motor imitation in a variety of circumstances.[28,29] As in the case of dolphins, which are perhaps the best combo vocal/motor imitators of all non-human animals, it is difficult to know how complex the cognitive mechanisms are that are responsible for imitative behaviors. But it is likely that a number of species have demonstrated a sense of agency via their capacity for imitation.

Self-awareness

In a groundbreaking article published in *Science* in 1970, Gordon G. Gallup Jr. introduced the concept of the mirror self-recognition test (MSR).[30] This test would go on to become the *de facto* standard for testing for self-awareness in non-human animals.[31] The test involves marking an animal's body with odorless paint or some other dye in an area that the animal is unable to see without the assistance of a mirror. If the animal, after being marked, proceeds to inspect the mark using its image reflected in the mirror, the animal is said to have passed the MSR test. Gallup's original article described two chimpanzees that passed the mark test, and two stump-tailed macaques and two rhesus monkeys that failed the test. It was the first evidence that an animal other than a human could recognize itself in a mirror, and fitted in well with the research at the time suggesting that chimpanzees might possess rich mental lives. Gallup concluded that "insofar as self-recognition of one's mirror image implies a concept of self, these data would seem to qualify as the first experimental demonstration of a self-concept in a subhuman form."[32]

As of the late 1980s and early 1990s, research was showing that dolphins might have cognitive abilities on par with that of chimpanzees in certain domains. Inspired by these findings, Diana Reiss teamed up with a student in Gallup's laboratory, Lori Marino, to perform the first ever test of MSR in dolphins.[33,34] In 1990, Reiss and Marino introduced a mirror into the pool of two bottlenose dolphins at Marine World in California. After a week of exposure to the mirror, the two dolphins (Pan and Delphi) began directing behaviors at the mirror that suggested they were inspecting their own reflections. This included contingency checking (e.g., bobbing their heads to see if the reflected image moved the same way), and self-directed behavior (e.g., examining the inside of their mouth and inspecting their tongues or other body parts that they cannot usually see without the aid of a mirror).[35] The dolphins were eventually marked with a waterproof dye in a similar fashion used by Gallup in his mark tests. Unfortunately, Pan and Delphi only seemed to examine the area where the mark had been made after the mark was removed (likely because the zinc oxide used as the mark could be felt by the dolphins, and thus they were quite agitated until the mark was removed),[36] which was insufficient evidence to conclude that they had passed the MSR test.

Soon after these experiments, Ken Marten at Project Delphis in Hawaii recorded similar examples of his dolphins inspecting themselves in mirrors, including examining marks researchers placed on their bodies.[37] However, this study was generally considered to be inconclusive evidence for MSR in dolphins by the scientific community.[38,39,40,41] A subsequent MSR study by Marten and Fabienne Delfour recorded both killer whales and false killer whales displaying contingency checking and self-directed behavior in front of the mirror.[42] On one occasion, one of the killer whales appeared to focus attention on the mark by rubbing the marked parts of her body against the side of the tank and returning to the mirror to inspect the area, presumably because she expected to see a change in the marked area. Unfortunately, this single occurrence of just one of the animals interacting with the mark was, while highly suggestive of the conclusion that killer whales pass the MSR, not the smoking gun that scientists craved. Part of the problem with the MSR as it applies to dolphins is the difficulty in interpreting their behavior as "inspecting the mark." Unlike primates—which have hands that can touch the mark—dolphins are only capable of twisting and turning to look at the mark (or possibly rubbing the mark on an object), which makes it hard for human observers to differentiate between self-directed behavior and mark inspection behavior.

Conclusive evidence that dolphins pass the MSR test was published by Reiss and Marino in 2001 following a second experiment they conducted with dolphins at the New York Aquarium.[43] In her book *The Dolphin in the Mirror,*[44] Reiss provides a fantastic account of the intricate study design and painstakingly crafted controls she and Marino implemented to be sure their results would be iron clad evidence of MSR in dolphins, as well as the quirky and frustrating process of scientific peer review. Unlike the previous attempts, this study unequivocally demonstrated the dolphins' repeated attempts to inspect the mark in the mirror by twisting and turning their bodies to bring the mark into view. Reiss and Marino concluded that this was evidence that the dolphins displayed self-recognition, but that "the question of whether dolphins are capable of more complex forms of self-awareness, such as introspection and mental state attribution, remains unanswered."[45]

At this point, you might have noticed that a number of different phrases have cropped up all describing related concepts: self-recognition, self-awareness, self-concept, concept of self, introspection, consciousness, mental state attribution, etc. In their 2001 article, Reiss and Marino described the dolphins as displaying self-recognition, but not self-awareness, and noted that "a provocative debate continues to rage about whether self-recognition in great apes implies that they are also capable of more abstract levels of self-awareness."[46]

This confusion as to what exactly the MSR test is telling us about the animal mind has been around since the day Gallup's original article was published. What are we talking about when we say "abstract levels of self-awareness" as opposed to "basic" or "low-level" self-awareness? Animals' experiences with MSR tests provide us a messy assemblage of successes and failures for different species, which show how difficult it is to interpret the "levels" of self-awareness responsible for behavior in front of a mirror. Aside from dolphins, all of the great apes, Asian elephants, and magpies have passed the MSR test. Capuchin monkeys don't pass the MSR test, but they appear to understand that their mirror image is not a "stranger" monkey, evidenced by the fact that they are friendlier to their reflection than they are to an unknown monkey.[47] Rhesus monkeys, which also fail the MSR test, have been observed using mirrors to inspect parts of their bodies they cannot see (i.e., self-directed behavior), which has led some researchers to conclude that they "recognize themselves in the mirror and, therefore, have some form of self-awareness."[48] B.F. Skinner, who was never quite convinced of the appropriateness of using self-awareness as the explanation for what Gallup's chimpanzees were doing in front of the mirror, published an article in *Science* in 1981 showing how he trained pigeons to peck at spots on their body which they could only see with the aid of a mirror.[49] Squid seem interested in their mirror reflections, and even tend to spend more time inspecting and touching their mirror image when marked on their heads with dye.[50] In one test to see how wild dolphins responded to mirrors, individuals actively avoided the mirror, ignored the mirror, or in one case, displayed aggressively at the mirror (as if it were another dolphin).[51] And while it has been generally accepted that humans first pass the MSR test between eighteen and twenty-four months of age, a study found that almost all children given the MSR test in Kenya—including children as old as six—failed to touch or inspect a mark (i.e., Post-it note) in the mirror upon encountering their reflection.[52] Instead, the children froze in front of the mirror and just stared at their reflection, seemingly in shock. While North American children usually passed the MSR test by inspecting the mark, far fewer children from other cultures pass the test (e.g., Fiji, Saint Lucia, Grenada, and Peru).

So what is going on here? Are these results suggesting that squid are more self-aware then Kenyan children? Obviously not. A possible explanation was that the Kenyan kids knew darn well that it was their reflection in the mirror, but that they live in a culture where standing out from the crowd and being regarded as a unique individual is not valued in the same way as it is in Western culture. Confronted with their own image in a mirror—an image showing them with a bizarre Post-it note stuck to their bodies—their cultural upbringing resulted in their being unable or unwilling to react the way the researchers expected.

Similar reasoning might explain why gorillas confused researchers for years by being the only great ape species to consistently fail the MSR test.[53] Gorillas are often described by researchers as being easily embarrassed, and seem acutely aware of who is watching them.[54] In early MSR tests in gorillas, one subject, Michael, appeared to notice a mark using the mirror made by an experimenter on his brow, but quickly turned away from the mirror and rubbed the mark off.[55] When he was later marked on his nose, he appeared to spot the mark in the mirror and requested (via his symbolic communication system) to have the lights turned off and the drapes closed in his enclosure. When this request was not granted, he retreated to the corner and rubbed the mark off of his nose. These are regarded as failures of the MSR test insofar as Michael did not inspect the mark in the mirror, but suggest that Michael was well aware that his nose was marked, and attempted to address the problem in private. Moreover, unlike chimpanzees and bonobos, eye contact is often regarded as a sign of aggression or a dominance display in gorillas, which might make it rather uncomfortable for a gorilla to stare at another gorilla in a mirror—even if they know it is their own reflection looking back at them.

These examples highlight one of the major drawbacks of the MSR test. Not all species—including humans—should be expected to react to being marked and looking in a mirror in the same way. MSR might work well for primates, which are species that engage in grooming behavior and are thus constantly seeking out parasites or detritus clinging to their bodies, but perhaps not so well for other species that do not groom. It's interesting to note that dolphins don't inspect the marks made on other dolphins as the great apes do during MSR tests, which is to be expected from a non-grooming species. And one of the explanations as to why elephants initially failed the MSR test was that they have a habit of covering themselves in debris (e.g., dirt), which might make a mark on their heads a far less salient cue for an elephant than for a species that likes to keep its skin clean. For dogs, which process much of the world via the sense of smell, recognition of others and self might occur primarily via olfaction. So while dogs consistently fail MSR, it has been suggested that they might yet be self-aware based on their ability to distinguish between their own and other dogs' scent markings (i.e., urine).[56] Similarly, it has been suggested that crickets might be self-aware after a study found that they are able to differentiate their own pheromone scent from that of other crickets.[57] Of course with these sorts of experiments, it's difficult to know where the animal is simply making a basic sensory discrimination (e.g., this scent versus that scent) as opposed to making a link to the idea of *my* scent.[58]

The question we're left with is: how useful can a test for self-recognition and self-awareness in animals really be if it can't reliably be used to uncover these abilities in humans—which we know with 100 percent certainty are self-aware? If MSR produces false-negatives, consider the vast number of species that might yet be self-aware but simply fail MSR because of idiosyncratic behavioral, ecological, or evolutionary issues—like having no interest in grooming, a fear of eye contact, or simply no patience for these kinds of annoying experiments.

But even if we ignore these objections and accept that MSR should remain the acid test for self-awareness in animals, we're still left with the problem of what kind of self-awareness MSR is capable of uncovering. Some have argued that MSR is nothing but a form of body-awareness, where the animal has simply learned that the visual input from the mirror correlates with the sensory motor information it is receiving from its body (i.e., kinesthetic–visual matching),[59] allowing it to use the mirror as an additional, novel form of sensory input for making the most basic of self-knowledge distinction: my body versus not my body.[60,61,62,63] This explanation does not require the animal to have any awareness of its own mind. Others have noted that MSR and empathy develop in humans at about the same age (i.e., eighteen to twenty-four months), and those animals that pass MSR (e.g., dolphins, elephants) are also often described as displaying empathetic behavior[64] or other behavior suggesting that they have an awareness of the minds and emotions of other beings. Thus, the human developmental evidence correlates with the phylogenetic evidence to suggest that MSR might be directly connected to—if not direct evidence of—the most complex/abstract form of mind awareness in animals. Consequently, some scholars have argued that whenever self-awareness appears in a species, it's almost certain that awareness of other minds has evolved with it.[65,66] Between these two extremes, we find a dazzling array of other explanations and a very passionate debate among scholars which leads to two oft-repeated conclusions: 1) there is likely to be a spectrum or continuum of self-awareness within the animal kingdom, and 2) the MSR test is not capable of saying where on this continuum an animal should be placed.

The fabled continuum of self awareness

Gallup did not initially discuss the possibility of a continuum of self-awareness that might lead to different responses to mirrors, but classified animals as either passing or failing MSR and thus having or not having a concept of self.[67] But the modern view of MSR suggests that different levels of self-awareness might be responsible for different kinds of responses in front of the mirror. For human development, psychologists typically have a good grasp of when different stages of self-awareness develop in children and how it correlates with their performance on MSR tests. But combing through the literature on MSR and self-awareness in animals makes one fact abundantly clear: there is no consensus as to how to define different kinds/levels of self-awareness in animals or how they relate to MSR behavior. This so-called continuum of self-awareness to which scholars refer is often a kind of placeholder for a confusing collection of cognitive and psychological processes that combine in unknown ways to produce unobservable and poorly defined types/kinds of awareness that might or might not be responsible for animals' performance on MSR tests. The following list highlights many of the cognitive/psychological concepts a self-aware animal displaying MSR might be experiencing.

- I have a body that I can feel
- My body is separate from other objects in the world
- I can move my body
- There is just one of me controlling my body
- When I move my body I can feel it moving
- When I move my body I can see (parts of) it moving
- I can feel parts of my body that I can also move, but cannot see
- I have control over my actions and can choose to move my body or not
- There are other things in the world with bodies that look just like my body, but they are not my body because I cannot move them
- When I do some things with my body, good or bad things can result (e.g., pain, pleasure)
- I like good things and do not like bad things
- I want to make my body perform actions that result in good things
- I am able to make a plan as to how to move my body to obtain good things without first needing to move my body
- I have feelings and desires that make me want to move my body in certain ways (e.g., anger, hunger)
- I can experience feelings and desires without needing to move my body

- I know which things I do or don't do that will result in these feelings and desires
- I am able to think about these feelings and desires, decide whether I like them or not, and can plan to act in ways that make them appear or disappear
- I know that I have existed and experienced feelings and desires in the past and will exist and experience feelings in the future
- Other things with bodies like mine move their bodies in similar ways to mine—they might be trying to do similar things to what I am trying to do
- These other things might have plans about how to move their bodies just like I do
- These other things might have feelings and desires that make them want to move their bodies like me
- These other things might know that I have feelings and desires that make me want to move my body
- These other things might be observing me and making guesses as to what I might want to do based on their ideas about what feelings and desires I might be experiencing.

It is not the case that the items on this list occur in some kind of definite order relating to stages of self-awareness (either developmentally or phylogenetically), which map directly only stages of MSR competence in animals. As we've seen in various disorders like body integrity identity disorder or schizophrenia, things that logically should be built in a hierarchical form (e.g., body-awareness followed by agency) can become disconnected from each other resulting in bizarre states of awareness (e.g., feeling like your leg is not your leg but still being able to move it). So it's entirely possible that animals—whose minds are still a black box when it comes to these issues—might possess unexpected and very un-human-like combinations of these cognitive/psychological concepts which create species-unique forms of self-awareness resulting in species-unique responses to mirrors.

When labeling the kinds of self-awareness dolphins might possess as evidenced by their performance on the MSR test, scholars and other pundits are rarely consistent in terminology. For example, Reiss and Marino did not explicitly state in their 2001 article that they had discovered self-awareness in dolphins, only that they had found self-recognition, which they implied was an indicator of self-awareness.[68] Marino would eventually state that the MSR study showed that dolphins "have a sense of self,"[69] and testified before the US Congress that "this and other studies show that dolphins have self-awareness, a sense of themselves not unlike our own."[70] Reiss subsequently used the terms "self-awareness" and "consciousness" to describe what the MSR study had found in her popular writings.[71] Thomas White suggests that the MSR test "requires" self-awareness,[72] and that dolphins "possess not simply consciousness but self-consciousness."[73] Leah Lemieux suggests that self-awareness and sentience (but not consciousness) can be revealed by MSR,[74] whereas Toni Frohoff suggests that the kind of self-awareness seen in MSR is just one "aspect of sentience."[75] A long list of other terms (e.g., autonoetic consciousness, reflective consciousness)[76] have been used to describe the kind of awareness dolphins display during MSR. Depending on the scholar, these terms could involve any combination of the cognitive/psychological concepts I listed, and more often than not, scholars simply don't define or describe the terms they use in enough detail to allow us to guess which concepts they might mean.

So what does Reiss and Marino's study confirming that dolphins pass the MSR really tell us about self-awareness in dolphins? At the bare minimum, a dolphin must know that the image in the mirror corresponds to its own body plan, strongly suggesting that dolphins have body-awareness and agency. Thus, if we equate self-awareness with the idea that a dolphin knows "it's my body there in that mirror," then dolphins are self-aware (or conscious, self-conscious, or sentient, depending on which vocabulary word strikes your fancy). But many scholars, like Gallup himself, suggest that "there is much more to being self-aware than merely recognizing yourself in a mirror."[77] What about these more abstract kinds of self-awareness that might involve a dolphin reflecting on its thoughts or pondering the thoughts of others—is that what the MSR study has shown for dolphins? This is impossible to say. A decade has passed since these results were first published, but the scientific community is not much closer to settling the debate on this question.

There are two major problems with the MSR test that are likely to keep the debate raging for decades to come. First, there might be plenty of animals that have the same level of self-awareness as dolphins that simply fail the MSR test because it's not the right test of self-awareness for them. So we can never state with certainty that an animal that fails the MSR test is either 1) incapable of recognizing an external representation of itself, or 2) not self-aware. Second, the MSR does not provide us any means of differentiating between an animal that possess basic body awareness, and an animal that possesses a human-like awareness of their thoughts, desires, emotions, and maybe the thoughts, desires, and emotions of others. Both kinds of animals will react the same way to their reflected image during the MSR test. All of this means that the MSR test provides us tantalizing clues as to the level of self-awareness dolphins might possess, but nothing in the way of clear answers.

Self-awareness: beyond the mirror

Mirror self recognition does not provide the only line of evidence that researchers suggest might indicate that dolphins have a sense of self. Lou Herman has argued that the studies showing that dolphins are able to understand labels of different parts of their body, use those body parts to follow specific instructions, and self-imitate their own behavior, strongly suggests that they possess a sense of both body-awareness and agency.[78] As we have seen with the MSR test, this level of body-awareness and agency often receives the label *self-awareness,* although Herman also used the terms *consciousness* and *self-consciousness* in his review of these topics. He also argues that dolphins' ability to imitate the actions of others (both humans and other dolphins) might mean not only that they are able to differentiate between themselves and others, but might attribute some sort of agency to other beings as well. Others have argued that signature whistles (which will be discussed in the section on dolphin language) are evidence of self-awareness.[79,80] If signature whistles are truly the equivalent of a name that a dolphin can use to both refer to itself and to others, then this would necessitate some form of self-awareness. Of course, it is still a matter of debate as to how dolphins really do use their signature whistles, or if they have any mental representation of themselves or other dolphins when they use them, so this is not an unproblematic argument for self-awareness.

The MSR provides fairly convincing evidence that dolphins have a sense of body-awareness, but does not do a very good job of telling us if dolphins are aware of their own thoughts. There are, however, other experimental designs that have given us evidence that dolphins are able to think about their own thinking. This ability, called *metacognition,* might be synonymous with what some scholars consider to be self-consciousness, introspection, or abstract self-awareness. Unlike these labels, however, metacognition has been distilled into easier-to-test-for operational definitions, with the original definition provided by John Flavell as "one's knowledge concerning one's own cognitive processes or anything related to them."[81] Cognitive processes involving the perception of stimuli and processing of information occurs in the brain at all times, but the mind is by no means aware of all information processing taking place. Metacognition is a skill that allows one to monitor these processes (to some extent) and to control them. It should be possible to have a sense of agency (e.g., I am aware that I can move my body) without necessarily having meta-knowledge of the thoughts that give rise to the movements. Metacognition is an extra layer of knowledge that allows the mind to represent representational states (e.g., I am aware of the fact that I am aware that I can move my body). This results in the "feeling of knowing,"[82] and the ability to judge one's own level of confidence as to whether or not one knows or is experiencing something, and adjust one's thoughts and behaviors accordingly. It might be a uniquely human property, and falls somewhere under the vague umbrella categories of "abstract" or "complex" self-awareness.

Testing for metacognition in animals began in the mid-1990s, spearheaded by University at Buffalo psychologist J. David Smith.[83] The first animal to become the subject of tests of animal metacognition was a dolphin named Natua at the Dolphin Research Center in Florida.[84] Smith and his research team began by asking Natua to discriminate between two categories of tones, which they divided arbitrarily between a high tone (2,100 Hz), and low tones (anything between 1,200–2,099 Hz). Natua learned this task easily, and had no trouble choosing between the two categories, receiving a fish reward for doing so. But as the low tones were raised in pitch so that they approached 2,085 Hz, Natua had a harder time differentiating them from the 2,100 Hz tone, and began making mistakes. Getting it wrong meant that Natua would receive a time-out period with no fish rewards. At this point, if Natua was experiencing something in the way of metacognition, he would have been aware that he was having a hard time distinguishing between two tones that were close in frequency. Smith's ingenious solution for testing for this was to introduce a response paddle that the dolphin could press that stood in for the concept of "uncertainty"—pressing this paddle would result in a short time delay followed by a new discrimination task featuring two very easy to distinguish tones, and thus an easier shot at getting a food reward. Obviously, the fastest way to a food reward was to choose the correct response, but if Natua wasn't sure what the correct response was, the next fastest was to choose the "uncertainty" paddle as opposed to making a guess and getting it wrong/having to wait. The results of this study showed that as Natua neared his threshold for distinguishing between the two tones and would otherwise have been performing close to chance, he began using the uncertainty response. Often in this situation he would noticeably hesitate, slowly approaching the two choice paddles, and sweeping his head from side to side as if he couldn't decide what the correct answer was. The interpretation of these results was that Natua knew that he was having a hard time deciding (by availing of some form of metacognition), and ultimately chose the uncertainty paddle when the task was too difficult.

Is this irrefutable evidence of metacognition in dolphins? Maybe. Researchers (including Smith) have suggested the possibility that Natua's responses might have been due to some sort of associative learning based on low-level cognition that does not involve metacognition,[85,86] and that the experimental set-up for the early tests of metacognition in dolphins does not provide as strong evidence as later tests involving great apes, macaques, and rats. However, many scholars have argued that the Natua experiment does provide convincing evidence of metacognition in dolphins.[87] Even if this is the case, it's not known if the quality of this metacognition is identical to that of humans, nor if this metacognition extends beyond monitoring anything other than decision making. As Smith himself pointed out, "one does not have to attribute full consciousness to the dolphin to explain its behavior"[88] in these experiments. Like the results of the MSR test, it's hard to say for certain what level of richness in metacognitive awareness dolphins might need to perform well in this experiment (let alone what exactly "full consciousness" is supposed to mean). Via an argument by analogy,[89] a dolphin pressing an uncertain response paddle is reflecting on its thoughts in a similar way to a human. But, as is always the case, this need not be the most parsimonious explanation, and we can't know for sure. Although these tests are generally considered a fairly strong indication of metacognition, we are stymied again by the question of what "levels" of metacognition or awareness are lurking in the dolphin mind.

The Minds of Others

Somewhere in my list of cognitive/psychological concepts that a self-aware animal might possess, you might have noticed that a fuzzy line had been crossed where mind awareness spills over from the awareness of one's own mind, to the awareness of other minds. Here too we find a spectrum of ill-defined awareness that starts with the understanding that other entities in this world might have a sense of agency, through to the understanding that they have knowledge and belief that drives their actions. An awareness of this sort is referred to as having a theory of mind,[90] or mind reading ability, and is one of the advanced cognitive skills that PETA and other activist groups have suggested has been demonstrated in dolphins. We've seen that some believe that self-awareness and theory of mind are intimately linked, and that animals that pass the MSR test or display metacognition are likely to also possess identical knowledge about the minds of other beings. However, of all the kinds of awareness we've discussed up to now, theory of mind is by far the hardest to prove in the lab, and the kind of awareness for which evidence from dolphin studies is just mildly suggestive.

As previously noted, Herman suggested that a dolphin's ability to imitate the movements of other dolphins and humans might suggest that they are able to differentiate between agency as displayed by themselves, and agency as displayed by others.[91] Agency in this sense might mean that a dolphin knows that these other entities are beings that have control and ownership over their actions. It does not tell us, however, the extent to which a dolphin attributes theory of mind to these other beings. Ferreting out the difference between attributing agency and attributing theory of mind is the central issue in this discussion. There are three ways that a dolphin could interpret the behavior of other entities it comes across in the world, corresponding with three levels of complexity in the understanding of the minds of others:

1. *Behavior reading*: The dolphin could understand that another entity displays changes in behavioral states that correspond with something akin to agency.
2. *Mental state attribution*: The dolphin could attribute mental states like "wanting" or "intending," as well as perceptual states like "seeing," "hearing," or "paying attention" to other entities.
3. *Theory of mind*: The dolphin could understand that other entities "know" or "believe" things.

Aside from Herman's studies of imitation, a number of other studies suggest that dolphins are adept at behavior reading, and possibly mental state attribution, but not necessarily theory of mind. Of course, all of these results must be taken with a grain of salt, as some scholars have argued that all experiments intended to sniff out the presence of animals' ability to attribute some kind of mental states to other entities that have been conducted up to now are fundamentally incapable of settling this question.[92] The so-called "logical problem" suggests that it is impossible to differentiate between an animal's ability to read mental states and an animal's ability to read behavior that is a result of these mental states. Thus, we do not currently know—and might never be able to know—if animals are true mind readers or simply skilled behavior readers.

Gaze

Dolphins have demonstrated the ability to spontaneously follow the direction of a human researcher's gaze to locate objects in an object-choice experiment.[93] Ake and Phoenix, two of the language-trained bottlenose dolphins from Kewalo Basin, were given instructions to carry out a behavior on one of two objects that was located on either side of the researcher. Instead of naming the object using Ake or Phoenix's artificial language system, the researcher gave the command and then looked over to the target object by turning his/her head. Without having received any prior training as to what this signal might mean, Ake and Phoenix understood that the direction of the researcher's gaze indicated which object to choose—something they understood from the very first test. Similar results have been obtained by other researchers with five other dolphins.[94] Ake and Phoenix were also able to understand the gaze even when it was not accompanied by the dynamic turn of the head (i.e., a static gaze revealed after lowering an opaque board that was previously obscuring the researcher). So it was likely the direction of the researcher's head as opposed to the movement that was the salient cue on these tasks.

Similar results from an experiment with rhesus monkeys have led researchers to suggest that the monkeys had displayed mental state attribution in the form of knowing that other animals/researchers are "seeing" or "perceiving" something upon which their gaze is focused.[95] Because dolphins, unlike primates, appear to understand gaze cues instantaneously (i.e., without needing repeated trails/exposure) it might well be evidence that they are adept at some form of mental state attribution insofar as associative learning could be ruled out. But this is perhaps not the best interpretation for these results. Dogs too are able to understand human gaze without any prior training,[96,97] with the possible explanation that dogs have been selectively bred for their sensitivity to human gaze, driven not by their ability to attribute mental states, but simply by an ability to read a human's behavioral state. Of course dolphins are not domesticated, which makes it an intriguing question as to why they seem skilled at following human gaze. It's still possible of course that Ake and Phoenix could have accidentally learned to follow human gaze based on their long-term exposure to humans, although experimenters took steps to avoid this.

Learning or instinctually knowing how to follow gaze is perhaps rather widespread in the animal kingdom, which makes it difficult to guess what is going on in the case of dolphins. Goats are able to use gaze cues from other goats to find food,[98] and tortoises are able to spontaneously follow the gaze of other tortoises.[99] Very young infants that have not yet developed theory of mind and possibly not even mental state attribution can also follow the gaze of an adult to an object of interest.[100] Unlike dolphins, some animals have been tested for their ability to follow a gaze cue around a barrier. All of the great apes as well as ravens are able to follow human gaze around a barrier to find a reward, which might constitute stronger evidence that they attribute mental states of "seeing" to the human researchers than unobstructed gaze following experiments.[101,102] Like most tests of this sort, it's all but impossible to differentiate between the three levels of complexity that might be driving gaze following, and debates rage in the literature as to what is the correct interpretation.[103] Regardless of the level of underlying complexity, dolphins stand out on these tests for their ability to understand human gaze cues so easily.

Pointing

Pointing is a universal human behavior usually involving extending a finger (sometimes together with the hand and arm) in the direction of an object or event of interest. Unlike simply extending the hand, which is a gesture seen in great apes as well as humans and is considered a request for an action or object, pointing is used by humans to direct another agent's attention. It develops in two stages, with imperative pointing developing at about twelve months of age, and used similarly to the great ape gesture to request an object/action. Imperative points are produced by infants to try to manipulate the behavior of an adult and are produced whether or not an adult is paying attention to the infant's behavior, suggesting that the infant has not yet attributed mental states or theory of mind to the adults. At twenty-four months of age, however, infants begin to use declarative points. These kinds of points are produced when the infant is aware of the gaze or attentional focus of the adult, and is actively trying to change their focus of attention. At this stage, it is possible that the infant is attributing mental states like "attention," or "seeing" to the adult, and quite possibly "beliefs." However, it is still possible that declarative pointing is only used to share or direct attention to distant objects or events and that it's not necessarily the case that the pointer attributes abstract mental states to the other entity. Thus for some scholars, both imperative and declarative pointing rely only on behavior reading, and not mental state attribution or theory of mind, whereas others believe declarative pointing involves mental state attribution and/or theory of mind.

Animals typically have a very difficult time understanding what a human pointing gesture means. A number of species have been subjected to experiments to test their ability to follow a point to an object in an object-choice task, with equivocal results. Gray seals learned to use a point to find a food reward, but it was likely they had simply learned to use the cue via operant conditioning, and did not understand that the point was intended to cue their attention.[104] Some monkey species appear to orient their attention upon seeing a pointing gesture (e.g., cotton top tamarins,[105] gibbons,[106] capuchin monkeys[107]) whereas others have not (e.g., rhesus monkeys,[108] lemurs[109]). But wildly different experimental methods—which sometimes involve producing a point in combination with a gaze cue—make it difficult to interpret the extent to which these species did or did not understand pointing. There is a heated debate as to whether or not chimpanzees and other great apes can understand declarative pointing.[110,111,112,113,114]

In contrast to these ambiguous results, dogs (but not wolves) are particularly skilled at understanding a human pointing gesture.[115,116] They can spontaneously follow points to both near and distantly located objects, and even do so from a very young age.[117,118,119,120] Similar to the explanation for dogs' ability to understand human gaze, the domestication process might have allowed humans to breed an animal that is very sensitive to human referential signals like pointing. The domestication process might also explain why both domestic goats and horses have shown some ability to understand human pointing.[121,122]

Only one other species has performed as well as dogs on tests of pointing comprehension: bottlenose dolphins. When the language-trained Kewalo Basin dolphins were first given the human pointing gesture by a human as an indication of the particular object on which they were being asked to perform an action, they responded with 81 percent accuracy.[123] These dolphins had been passively exposed to points in various contexts before, but had never been explicitly trained on the "meaning" of the pointing gesture. Follow-up experiments showed that these dolphins could understand different forms of the pointing gesture to both near and distant objects, including fully extended arm and finger points (ipsilateral points), exaggerated points (including movement of the body in the direction of the point), and cross-body points given as either dynamic or static points.[124,125,126] The dolphins could even understand a command sequence consisting of two points referring to two separate objects (e.g., take THIS ball in THAT basket). No species other than humans and dolphins have shown an ability to understand two separate points in the same command sentence. Also remarkable is the fact that dolphins could understand momentary points[127] that were held only briefly, as opposed to the long-held static points that chimpanzees have such difficulty with. Because of the immediate high levels of success the dolphins had understanding all of these kinds of pointing behaviors, it is unlikely that they had the time to use associative learning or operant conditioning as a means of learning the gesture's meaning. Alain Tschudin and colleagues replicated these findings with a group of six dolphins that had had no prior exposure to the human pointing gesture, lending further weight to the idea that dolphins spontaneously and easily understand the meaning of human pointing.[128] Why dolphins—an animal that has no arms, hands, fingers, or any other appendage that could produce something resembling the human pointing gesture—should have this ability is still a mystery. It is possibly related to their ability to follow the direction of each other's echolocation beams in the wild, making them acutely aware of each other's direction of attention.[129]

Is dolphin pointing comprehensive evidence that dolphins are attributing mental states to humans, or have a theory of mind? Probably not.

Of course, in adult humans, point comprehension goes hand in hand with these two forms of complex understanding of other's minds. But the evidence is still shaky as to the extent to which animals ever develop similar levels of understanding. Even scholars who disagree as to whether evidence exists that animals are capable of mental state attribution still appear to agree that tests of point comprehension are incapable of differentiating between behavior reading and mental state attribution. Many people might have a gut feeling that dolphins understand pointing because they really do know what humans "want" when they point to something. But I suspect fewer would say the same about dogs, even though both species understand pointing in similarly impressive ways. And chimpanzees perform so well on related tests (e.g., MSR, gaze following, metacognition) that it would seem strange to conclude that their lack of comprehension of pointing was due to mentalization skills that are less complex than that of dolphins or dogs. Point comprehension is yet another test that leaves us with more questions than answers.

Moving beyond comprehension, dolphins have shown aptitude in their ability to produce pointing gestures. Whereas the evidence that great apes and other primates produce declarative pointing gestures is controversial, dolphins, an animal that doesn't even have hands with which to point, are first-class point producers. Six months after beginning a training program in which two dolphins, Bob and Toby at Disney's The Living Seas, were learning to associate symbols on an underwater keyboard with locations of objects/rewards in their tank (e.g., food, toys, and tools), the two dolphins began pointing to the location where they expected their food reward to be found.[130] This pointing behavior took two forms: 1) positioning themselves close to the object (i.e., within one dolphin body length) and remaining perfectly still with head and body in alignment with the food reward, and 2) alternating between this position and looking back at the diver who was approaching to retrieve the food reward. These are unusual behaviors for dolphins, and likely not a product of the dolphin inspecting the food reward (which would likely involve body movement or a sweeping head gesture as they inspected the food reward with echolocation). These behaviors were considered a spontaneous act because 1) the dolphins were not specifically reinforced for engaging in pointing behaviors, 2) were given the food reward regardless of their production of pointing gestures, 3) were never trained to produce pointing gestures, and 4) were not always rewarded with food when engaging in pointing gestures. Moreover, they produced these pointing gestures only when the humans were far away, and did not produce them if no humans were around.

It certainly seemed as if the dolphins were aware of the attentional states of the divers, which led researchers to develop an experiment to specifically test this hypothesis—this time using a food reward as motivation.[131] A divider was placed in Bob and Toby's tank that allowed them to inspect two jars (one of which contained a food reward) using either vision or echolocation, but blocked them from actually reaching the jars. The researcher then waited for the dolphins to produce some sort of point-like gesture to indicate which jar the diver was meant to open. The divers wore a mask that obscured their eyes, so it was assumed that the dolphins were able to guess the divers' direction of attention based on which direction their bodies/heads were facing. The dolphins produced far more pointing gestures when the divers were facing the dolphins (i.e., attending to the dolphins) than when the divers were facing another direction or actively swimming away from the dolphins. This suggested that the dolphins were aware of where the divers' attention was focused, which could indicate that they had attributed the mental states of "seeing" to the divers. Of course, as always, the dolphins could have learned an associative rule like "only produce these pointing behaviors when the diver is positioned such-and-such a way otherwise there will be no food," or otherwise have based its decision to point based on behavioral reading. But the spontaneous nature of their initial production of pointing gestures means this is a less likely scenario, and dolphin point production/comprehension might be evidence of mental state attribution.

Anecdotal evidence of dolphins pointing in the wild—aligning their bodies to "point" at the dead body of a conspecific—has been observed,[132] but it's still unclear if wild dolphins are attributing mental states like seeing or knowing to each other, or are just reacting to behavioral states without any idea that other entities possess minds like theirs. And it is still unclear if dolphins attribute mental states like "wanting" or "intending" to humans or other dolphins that are producing pointing gestures, and whether or not these would constitute something akin to mental state attribution. On one hand, some have argued that attributing intentions to other agents is proof that an animal holds an abstract understanding of other minds, whereas others have argued that attributing intentions is part and parcel of low-level behavior reading that has nothing to do with other minds whatsoever. Like everything in this discussion, scholars disagree on the different kinds of "intentionality" animals might attribute to other agents, with "low-level" intentions attributed to any object/animal that moves of its own accord (similar to agency), and "rich" intentions attributed to objects/animals that move of their own accord because they are driven by desires and beliefs. Unfortunately, research into dolphin pointing does not provide us with a smoking gun showing that dolphins engage in theory of mind, but does yield evidence suggestive of mental state attribution.

False belief

The vital difference between attributing mental states to others and having a theory of mind is the understanding that other entities can have *epistemic mental states*, which consist of things like believing, knowing, guessing, imagining, and pretending.[133,134] These states are considered more complex and abstract than mental states like seeing or wanting insofar as they might require an animal to reflect upon their own thoughts (via metacognition) and assume that other entities are also reflecting on their own thoughts via metacognition. This allows an animal to form a representation of another animal's mind that is separate from a representation of that animal's intentions. It has been suggested that human children are unable to attain this level of reflective thought until they are three to four years old,[135] and that deficits in theory of mind are what lead to autism, a disorder that has also been referred to as "mindblindness."[136] Some definitions of theory of mind might lump "intentions" or even "seeing" in with the other epistemic mental states, which blends theory of mind into mental state attribution as I have outlined it here.[137] Some scholars use the term "full-blown theory of mind" or "higher-order theory of mind" to denote the complex/abstract epistemic mental states I've listed,[138,139,140] although even this phrase sometimes lumps "intentions" in as well.[141] But theory of mind as I am using the term here is equivalent to "belief attribution,"[142] with attribution of intentions and other mental states being relegated to the category of mental state attribution (what others might call "minimal theory of mind"[143]).

Evidence for theory of mind in non-human animals is controversial. In a recent study, chimpanzees were shown to produce alarm calls more frequently in the presence of other chimpanzees who had not yet seen the "danger" (in this case a rubber snake placed by the researchers). The researchers concluded that these results imply that "chimpanzees keep track of information available to receivers and intentionally inform those who lack certain knowledge."[144] This is an intriguing finding that adds fuel to the debate as to whether chimpanzees possess theory of mind. Many researchers suggest that chimpanzees understand the intentions, goals, and perceptual states (e.g., seeing) of other animals, but fall short of displaying full understanding of epistemic mental states.[145] Some studies suggest that chimpanzees might keep track of what other chimpanzees "know" about food rewards that they've just seen,[146] although it's difficult to separate "knowing" from "intending" for these kinds of experiments.[147] Consequently, a debate rages as to if these and similar studies are convincing evidence of theory of mind for chimpanzees.

The most widely cited experimental designed claimed to be a direct test of theory of mind in animals is the false-belief task. The false-belief task tests an animal's ability to understand that another entity might have knowledge or a belief that might not correlate with reality, and was hailed as an empirical solution to the "logical problem" of differentiating between epistemic mental states attribution and behavior reading. When administered to human children in the form of the Sally–Anne test,[148] the child observes two puppets (Sally and Anne) both "watching" a marble being hidden under a basket. Sally then leaves the room (chaperoned by a researcher), and Anne moves the marble from the basket to another hidden location. When Sally returns, the child is asked where Sally will look for the object. If the child understands that Sally did not see the object being moved and that Sally holds the "false belief" that the object is still located in the basket, the child will say "Sally will look in the basket." This is very strong evidence that the child has a mental representation of Sally's mental states of "knowing" and "believing." Chimpanzees are unable to pass false-belief tasks that are based on this experimental paradigm.[149]

In a variation on this experimental paradigm, four dolphins at Sea World in Durban, South Africa were given the false-belief test in a pilot study orchestrated by Alain Tschudin in 1999.[150] The results of this study, which were published in Tschudin's PhD thesis but not the peer-reviewed literature, stem from an experimental design in which a dolphin was positioned between two empty opaque boxes placed at the side of their pool, one of which was baited by an experimenter. A second experimenter, called the "communicator," would then tap on the box containing the fish reward. A screen was placed between the dolphin and the first experimenter so that the dolphin was unable to see which box was being baited, but could continue to see the communicator. The communicator could see both the dolphin and the boxes, and would tap the correct box after the screen was removed. The dolphin then aligned his/her head with the box containing the fish as indicated by the communicator after the screen was dropped, and would receive a food reward. In the test for false belief, the communicator left the experimental area after the fish was placed in the box, and the first experimenter switched the position of the two boxes in full view of the dolphin. The communicator then returned and tapped on the box in the position where he/she had initially seen the fish placed. Of course, this was the wrong box, so if the dolphin understood that the communicator had a false belief about the location of the fish insofar as the dolphin had witnessed the boxes being switched, the dolphin should ignore the communicator's signal and choose the other box. Tschudin reported that the dolphins involved in this experiment did exactly this, performing above chance level, and thus ostensibly passing the false-belief test.

But this was not an unproblematic conclusion. According to Tschudin, it is possible that the dolphins solved this task by either 1) learning a simple rule that whenever the boxes are switched, the correct box was always the opposite of whatever box the communicator chose, or 2) since the first experimenter always knew the location of the fish, he/she might have inadvertently cued the dolphin as to the correct location. Tschudin published the results of a follow-up experiment that was intended to address these concerns, although these results appear not in the peer-reviewed literature, but in the form of a book chapter. To combat the problem of the dolphin learning a simple rule about which box to choose after they have been switched, additional trials in which the boxes were switched in the presence of the communicator were included. Also, a third experimenter—unaware of which box contained the fish—was brought in during the experiment when it came time to display the boxes to the dolphin, eliminating the possibility that he/she could inadvertently cue the dolphin to the location of the fish. With these controls in place, the dolphins performed similarly to the pilot study, with the noted exception that no individual dolphin had a performance that was significantly different to chance—only when the performance of all dolphins was pooled did the dolphins' performance attain above-chance ratings.

An important problem remained with this updated version of the experiment; namely that the communicator was still aware of which two boxes contained the fish insofar as he/she always knew when the boxes were being switched. So the dolphin still could have picked up on some sort of cue from the communicator other than the tapping of the box. A subsequent experiment with a new (i.e., naïve to the experimental task) dolphin and a control in place to compensate for this confounding factor was conducted, but the dolphin involved in this new experiment never managed to pass the initial phases of the training, and the experiment was scrapped.

Major problems with this experimental design remain, which is perhaps why these experiments should not be considered proof that dolphins pass the false-belief task. As Tschudin noted, the same four dolphins used in the pilot study were involved in the second study, so if they truly had used associative learning cues to succeed at the task during the first study, then this knowledge might have spilled over into the new experiment and been responsible for their success. Also, a successful performance required the researchers to pool the data from all dolphins as opposed to looking at individual performance, which might either be a symptom of small sample size, or could be interpreted as weak evidence of an effect. Also, it is entirely possible that dolphins are able to differentiate between a box containing a fish and an empty box based solely on the differences in the sound that a tapping finger makes for each type of box. Given dolphins' known skill with this type of auditory discrimination, this is a likely explanation as to how they could reliably choose between a baited and an unbaited box, regardless of what any human experimenters might believe or know about the location of the fish. Even if controls were put in place for these issues, this particular experimental design is incapable of eliminating the possibility that dolphins use the communicator's line of sight (or break in line of sight) as cue to which box they will tap, and thus is not an adequate test to overcome the logical problem and allow us to differentiate between mind reading/theory of mind (e.g., the communicator believes the fish is there) and behavior reading (e.g., the communicator's line of sight was last aimed at this/that box, which is why they will tap there).[151] Perhaps because of the many possible confounding factors listed by Tschudin and others, it's understandable why these experiments have yet to make their way into the peer-reviewed literature, where, as we learned from the saga of Reiss and Marino's MSR experiments, experimental designs must be iron clad if they are to get past the gatekeepers of academic journal publishing.

Only one experiment from the peer-reviewed literature has been proclaimed to be a direct test of theory of mind in dolphins.[152] In this experiment, captive bottlenose dolphins were given gestural commands by their trainers and tested to see if the attentional state of the trainer—defined as the direction that their head/eyes were facing—would affect the dolphins' ability to comprehend the signals. To this end, the trainers either rotated their body and/or head when giving the commands, wore a bucket on their head that obscured their eyes, or two trainers stood next to/in front of each other when giving the commands. In all cases, the dolphins' ability to understand the gestural commands was unaffected except when the trainers' bodies were facing away from the dolphins, thus obscuring the gestures. The position of the trainers' heads and the wearing of the bucket did not affect performance, suggesting that the dolphins were not cued to or interested in the trainers' mental/attentional states. These findings seem to contrast with previous research suggesting that dolphins are sensitive to the attention states of humans. The authors suggested that the particulars of the dolphins' training situation might have predisposed them to ignore head/eye cues from their trainers for this particular experiment. In any event, this particular experiment did not yield convincing evidence of either the presence or absence of theory of mind in dolphins.

Taken together, the evidence from tests of dolphins' ability to follow gaze, pointing gestures, and false-belief tasks do not provide us with a clear picture of how dolphins might conceive of the minds of other beings. Like the evidence from chimpanzees and other animals, the results are open to interpretation, with no compelling evidence that dolphins—like any other animal other than human beings—are capable of theory of mind. Looking at the totality of research conducted on this subject with other species, the evidence for mental state attribution for dolphins is weaker than it is for animals like scrub jays, for which researchers have been collecting an impressive body of evidence suggesting that jays make decisions on what foods to cache or retrieve based on whether or not other jays can see or hear what they are doing.[153,154,155,156] Even those critics who state that these jay experiments fail to demonstrate that the jays were engaging in mind reading as opposed to behavior reading suggest that this line of evidence is the strongest to date that an animal might understand something about the minds of others.[157]

Emotions

Do dolphins have emotions? Most definitely.[158] In fact, it's likely that all animals have emotions. It used to be the case that scientists believed that the study of emotion in animals was impossible from an empirical perspective. B. F. Skinner, the father of behaviorism, stated that "'emotions' are excellent examples of the fictional causes to which we commonly attribute behavior."[159] According to Skinner, it's simply not justifiable to suggest that it is fear or panic that is causing a dolphin to thrash wildly when trapped in a tuna net. Positing the existence of unobservable emotions as the driving mechanism behind observable behavior has long been a controversial practice in science. But this approach was forged in a time before science had the technology to peek under the hood, and study not just observable behavior of an animal, but the brain systems that give rise to it. Decades of research concerning the neurological underpinnings of emotion in both humans and animals provide solid ground for the belief that a dolphin trapped in a net is most certainly scared out of its gourd.

From our discussion of the dolphin brain, we were obliged to conclude that it is nigh impossible to suggest that specific brain properties and brain morphology are convincing evidence of the existence of things such as self-awareness, theory of mind, etc. These cognitive skills (most likely) rely on the function of the cortex, which varies considerably from species to species, providing few universal correlations between structure and function. However, when it comes to the idea of basic emotion,[160] we are facing a very different scenario. The primary function of brains is to oblige an organism to find food, stay alive (usually), and mate. In order to make these things happen, all brains have a built in means of arousing themselves to engage in activities vital to their survival,[161] including reacting appropriately to emergency situations—like being chased by a predator. These arousal systems are vital to the function of any organism. As an organism interacts with the world, these arousal states interact with higher brain regions resulting in anything from basic learning to complex cognition. Such states are, in vertebrates at least, generated by ancient subcortical regions that are nearly identical for all species. These arousal states are, by any other name, the basic emotions of fear, pleasure, pain, anger, disgust, and sadness, as well as sensations like hunger, thirst, sexual desire, etc. The argument for emotion in animals states that since all vertebrates possess nearly identical subcortical regions known to be responsible for mediating arousal states (i.e., the limbic system),[162] we can conclude that basic emotions are universal in these animals. There is, as always, much debate in the scientific literature as to which emotions should be classified as the basic emotions, and how they differ from sensory affects, complex emotions, and/or how universal these might be.[163] Nonetheless, the role of the limbic system in vertebrates is to create these emotions in the brain, and this appears to be a universal truth.

So, when we observe a dolphin thrashing in a tuna net as it's slowly hauled out of the water, we don't commit any scientific sins by suggesting that its brain is almost certainly working hard to produce the cocktail of neurochemicals necessary for the emotion of "fear" to help generate its thrashing behavior. We could test the dolphin's blood for fear/stress hormones, or stick its brain in a functional magnetic resonance imaging scanner and confirm quite easily that this is the case. Except for perhaps some size discrepancies, the non-cortical structures of the dolphin limbic system (i.e., the structures shared by all vertebrates, like the amygdala) are no different to other vertebrates, which suggests that they are playing their part in triggering fear in the dolphin brain.

But the elephant in the room is the problem of "subjective experience" of that fear emotion. Is the dolphin *aware* of its fear? Even those who argue for the universality of emotion in animals suggest that limbic system-generated emotions do not necessarily require anything in the way of subjective experience (presumably generated in the cortex) to efficiently rouse the brain to action. A vast number of human behaviors are driven by emotions (even complex emotions) of which we humans are not aware/conscious, affecting everything from our choice in mates to our preference in soda.[164,165] Consider cases of children with hydranencephaly—a disorder in which they are born without any cerebral cortex whatsoever.[166] Yet these children's brains are producing emotions that result in their smiling, crying, and laughing even though they lack the cortical brain structures that should make them subjectively aware of these emotions. So even using humans as an example, it's possible that an animal's brain could be riddled with emotion and yet have no awareness of it. This is where the controversy of emotions in animals (including dolphins) really begins. There are two unanswered questions in this debate: 1) how aware are animals of their own emotional states, and 2) do animals experience complex emotions like jealousy, grief, or empathy?

Subjective experience

Evolutionary biologist Marc Bekoff and ethologist Jonathan Balcombe have written extensively about the subject of emotion in animals.[167,168]

Their ideas are sometimes met with resistance in the scientific community[169] as their arguments concerning what animals are likely to be experiencing emotionally is not always supported by the kinds of empirical data that skeptical scientists demand. Controversies aside, scholars like Bekoff and Balcombe make a very important point regarding animals' subjective experiences of their emotions, which I will paraphrase here. Given that emotion in animals is likely universal (as far as basic emotions and arousal go), and given that human beings are known to have subjective experience of their emotions, what justification do we have for suggesting that animals do not experience their emotions in the same/similar way as humans? Especially if this explanation might better explain the many complex behaviors we observe in animals. Should not the burden of proof be on those who hypothesize that animals do *not* experience emotions? From the perspective of science, it's possible to argue that Occam's Razor is most properly applied in suggesting that dolphins do—as opposed to do not—have subjective experiences of their emotions. The presence of subjective experience of emotions is no more extraordinary than the suggestion of non-presence, so all things being equal, why assume non-human animals are different in this regard? This is, in fact, the same stance that humans apply to the subjective experiences of other humans (which, but for our ability to describe our experiences to each other with language, are just as inaccessible as animals' subjective experiences).

This approach provides us with a framework for interpreting the overwhelming scientific uncertainty as to how animals experience their emotions. But this is an almost philosophical approach to the problem that does not provide us with a clear-cut methodology for tracking down empirical evidence of the existence (or non-existence) of the subjective experience of emotion in animals (nor is it intended to). In their writings on this subject, Bekoff and Balcombe provide numerous examples of animals behaving in ways that *suggest* that they do experience emotions.[170,171] For dolphins, this evidence often takes the form of a human interpreting observations of dolphin behavior: a dolphin mother swimming "slowly and despondently," grieving for her lost calf;[172] two young killer whales holding a "lonely vigil" near the beach where their mother had died;[173] and a bottlenose dolphin with a "twinkle" in his eye which signaled his frustration.[174] These examples rely heavily on the common sense idea that if these animals are behaving in such a way that suggests they are experiencing emotion, we should assume they are. But it is always possible to object to these accounts as being empirically shallow insofar as they rely on assumptions, not direct tests for the subjective experience of emotions. Moreover, they might rely too heavily on their resemblance to human behavior, which need not necessarily be the prototype against which emotion-driven behavior be measured. If a cat does not act as if it's grief-stricken in a way that is recognizable to a human, should we then conclude that it is not experiencing grief? For all we know a cricket perched on a log looking at the setting sun is experiencing a cavalcade of emotion that does not result in any outward change in its behavior, and thus provides us with no recognizable (to a human) cues as to what's going on in its mind.

The so-called "argument-by-analogy" approach to emotion in animals (i.e., if the behaviors of an animal resembles that of a human when experiencing an emotion, we should assume the presence of human-like emotional experiences in the animal) remains quite controversial.[175,176,177] By all accounts, dolphins, like all mammals, behave in ways that suggest that they are experiencing basic emotions in similar ways to humans via the argument-by-analogy approach. And there are excellent philosophical and ethical reasons to want to adopt this stance.

But questions of ethics aside, I believe I am correct in stating that the current scientific consensus is that science is simply unable to tell us if animals are subjectively/consciously experiencing emotions in a similar way to humans, and that this remains part of the so-called "hard problem" of conscious experience.[178] The zoologist Marian Stamp Dawkins suggests that "it is important to be clear where observable facts about behavior and physiology end and assumptions about subjective experiences in other species begin. However plausible the assumption that other species have conscious experiences somewhat like ours is, that assumption cannot be tested in the same way that we can test theories about behavior, hormones or brain activity."[179]

There are, however, a handful of brain scientists who suggests that perhaps most of the scientific community is going about the problem all wrong.[180] No need for agnostic caveats or an argument by analogy; we already have unequivocal evidence from neurobiology that animals *do* experience basic emotions. This is the position taken by, among others, Jaak Panksepp, a neuroscientist who coined the term "affective neuroscience,"[181] and is perhaps best known for discovering that rats laugh when tickled.[182] Panksepp's research[183] suggests that a set of fundamental emotions (e.g., fear, lust, panic) is generated in the most ancient parts of the brain (primarily the hypothalamus) which is structured almost identically across mammalian species. He argues that these emotions do not require the presence of more complicated cortical structures, or even the basal ganglia or amygdalae (mid-brain structures), in order to be *experienced* by the animal. This is evidenced by the fact that an animal (like a rat) that has specific parts of the hypothalamus electrically stimulated will both act as if it is experiencing the respective emotion (e.g., fear), and "report" that it is experiencing that emotion by pressing on a lever that either stops (in the case of negative emotions) or continues (in the case of positive emotions) the stimulation. Both humans and other animals behave similarly during this type of experiment, with humans having the added ability to report their experiences using language.

In July 2012, a group of prominent brain scientists including Panksepp, Philip Low, David Edelman, Bruno Van Swinderen, Christof Koch, and dolphin scientist Diana Reiss signed the Cambridge Declaration on Consciousness in Non-Human Animals, which stated that most animals (including dolphins, insects, and octopuses) possess the neurological structures necessary to experience emotion (and the subjective experience of emotion), and that "the weight of evidence indicates that humans are not unique in possessing the neurological substrates that generate consciousness."[184] Of course, not everyone is convinced that Panskepp's research is irrefutable proof of subjective experience of emotions in animals. It remains to be seen if a neurobiological approach to the problem of emotion in animals will break through the epistemological barrier of the subjective experience problem and change the nature of the debate, or if we are going to be stuck with the "we'll never know for sure" caveat indefinitely. Regardless, the present scientific consensus is that dolphins almost certainly possess basic emotions, but we do not know (and might never know) the extent to which they experience them subjectively.

Complex emotions

Given how difficult it is to prove empirically that dolphins subjectively experience basic emotions like pleasure or anger, it's understandable why things become even more contentious when talking about complex emotions like empathy, jealousy, embarrassment, etc. Not only would these emotions need to be subjectively experienced to be homologous to their human-like counterparts, but they are predicated on the idea that the animal is self-aware and/or has something akin to an awareness of the minds, intentions, and beliefs of others. Marc Hauser summarized what is probably the general scientific consensus on the issue of complex emotions in animals at present, which takes all of these factors into account: "Emotions prepare all organisms for action, for approaching good things and avoiding bad things. But when we step away from the core emotions such as anger and fear that all animals are likely to share, we find emotions such as guilt, embarrassment, and shame that depend critically on a sense of self and others. I will argue that these emotions are perhaps uniquely human, and provide us with a moral sense that no animal is likely to attain."[185]

Many scientists studying dolphins have a gut feeling that dolphins are capable of complex emotions like empathy, but from an empirical perspective, direct evidence for this remains weak. I will focus here on a discussion of empathy, as this is the most commonly cited complex emotion that dolphins purportedly display. The argument in favor of the existence of empathy in dolphins comes from both observations of behavior as well as brain structure. Anecdotal accounts tell of pouting dolphins being consoled by their friends,[186] or dolphins grieving over the death of a loved one.[187,188,189] Female dolphins adopting orphaned calves might be a sign that they empathize with the calves' plight.[190] Stories of dolphins rescuing drowning swimmers are commonplace, a behavior that is purported to be driven by the dolphins' empathy for a floundering human swimmer. There are also plenty of observations of dolphins coming to the aid of injured or sick conspecifics, or even other species. These accounts of altruistic and epimeletic (i.e., care-giving) behavior in dolphins feed directly into the study of evolutionary biology, and particularly sociobiology, which attempts to understand why prosocial behavior (including altruism) evolved in animals.[191,192]

There is a continuum of prosocial behavior in the animal kingdom, which likely reflects a continuum of underlying causes ranging from the genetic/instinctual to the truly empathic. Social insects, like some bees and ants, will sacrifice their own wellbeing (and their lives) for the good of the colony, a behavior that is best explained by the genetic-relatedness of the individuals within the colony where sisters are more closely related to each other than their potential offspring. Thus, there is no need to posit complex cognition for prosocial ant behavior. There is ample evidence that many animals will display vicarious arousal when watching another of their species in distress, including mice showing a pain response when watching other mice in pain,[193] or mother hens showing increased heart rate when watching their chicks in uncomfortable situations.[194] This (widespread) behavior, termed "emotional contagion,"[195] might have evolutionary advantages in that an individual that is frightened or alarmed by the distress of others is more likely to flee (thus saving their own skin). Emotional contagion might also result in offering comfort to the distressed party which, in the event they are closely related, has a clear evolutionary advantage. These empathetic-like behaviors might have evolved without the need for something akin to self-awareness, a self-other distinction, or theory of mind. At the extreme end of the continuum we find humans, who are capable of committing acts of altruism that seem rooted entirely in a concern for others, with a full understanding that other humans experience joy or suffering in similar ways, and are not always explainable by genetic relatedness. The question is, are the anecdotes and accounts of dolphin altruism or epimeletic behavior better explained by underlying mechanisms that resemble human-like empathy, or something more akin to emotional contagion?

The difference between true empathy (sometimes called *cognitive empathy*)[196] and emotional contagion (sometimes called *primitive empathy*)[197] is often a question of whether or not that animal is simply responding to the observable behavior of others, or truly has an understanding that the other is experiencing an emotion (i.e., via theory of mind or perspective taking). Even the best designed experiments are not able to differentiate between these two options. In one recent experiment appearing in the journal *Science*, a rat was placed in a small restrainer device.[198] Its cage mate was visibly distressed by this situation, and struggled to free the trapped rat, delaying retrieving a chocolate reward so as to first help the restrained rat, and even going so far as to share the chocolate with its formerly trapped companion (an unexpectedly selfless act for a rat). This might suggest that the rats were motivated to help the trapped rats not simply because of emotional contagion, but by feelings of true empathy. But a dispassionate view of the situation might suggest that the helper rat freed the trapped rat in order to put an end to the trapped rat's distress calls or otherwise agitated state which, via emotional contagion, were causing the helper rat to experience stress.

Any account of an animal (including humans) providing aid to another could be explained by reciprocal altruism, whereby an animal might reduce its own fitness (temporarily) in order to aid another in exchange for receiving help from the other animal in the future. Reciprocal altruism, which has been observed in dolphins in various forms,[199,200] is found throughout the animal kingdom, and does not necessarily require an awareness of the minds of others in order to have evolved.[201] But, as Frans de Waal and others have argued, true empathy stemming from theory of mind could well be a reasonable (if not more parsimonious) explanation for altruism/reciprocal altruism in animals like chimpanzees and dolphins,[202,203] and possibly in the trapped rat experiment.[204]

Defining empathy, classifying empathy-driven behaviors, and discerning both proximate and ultimate causes of empathy-like behaviors in animals is a complicated and contentious exercise.[205] As is often the case, there is a broad spectrum of behaviors that fall under the vague category of empathy displayed by many/most animal species, and deciding whether or not a specific example (e.g., rats helping their cage mates) should be classified as true empathy or not is often a matter of either definition or taste. When it comes to the question of dolphins, however, many who argue that dolphins deserve extra legal protection cite empathy alongside other higher-order cognitive functions as a primary reason for special consideration.[206] This argument is not referring to empathetic behavior like emotional contagion, which is certainly not unique to dolphins, but instead is suggesting that dolphin empathy is complex—akin to the kind of empathy displayed by human beings. For humans, we know that this type of empathy is driven by theory of mind. Unfortunately, while dolphins might indeed have some form of self-awareness, the evidence that they possess theory of mind is, as we have seen, currently contentious and/or lacking. Consequently, we must accept that there is no hard evidence for true empathy (i.e., human-like empathy) in dolphins. We could only arrive at the conclusion that dolphins are capable of true empathy via argument by analogy (i.e., if the dolphin behaves as if it is experiencing empathy then it is experiencing empathy) which, while possibly a valid approach, is not an iron clad argument. Consider the added problem that an animal might experience empathy solely as a representation in the mind, with no physiological signs (e.g., increased heart rate) or action being taken (e.g., aiding a friend in distress). Thus, empathy and theory of mind might remain unobservable in some animals, which means an argument by analogy would yield a false-negative for species that don't act like empathic humans. The argument-by-analogy approach to empathy has obvious limitations as both a scientific tool, and an ethical argument.

This is often where an argument for true empathy in dolphins turns to neuroanatomy for hard evidence. For example, mirror neurons, which fire in response to an animal both performing and/or observing an action in others, are often touted as a neuroanatomical basis for empathy[207] (and maybe theory of mind)[208] in humans and other animals. While some have suggested that mirror neurons have been found in the dolphin brain,[209] there is in fact no evidence that this is the case. While it's not unreasonable to think that dolphins (like many animals) possess mirror neurons, confirmation of this would likely involve the same kind of invasive cortical experiments (i.e., placing electrodes directly in the brain) that led to its discovery in monkeys, which is no longer performed on dolphins in the United States and many other countries. In any event, current research suggests that mirror neurons might not be the primary means by which the brain understands the actions of others, or that they are otherwise responsible for empathy.[210,211] Some suggest that VENs are direct evidence of dolphins' ability for empathy,[212,213] but as has been discussed in detail in the chapter on the dolphin brain, this is not the case. The "well-developed" structure and position of the cortical areas of the dolphin limbic system (e.g., the paralimbic cortex, cingulate and insular cortex) have been suggested by some to be evidence that dolphins experience emotions more deeply than humans, and might be directly responsible for empathy and other complex cognition.[214,215,216,217,218,219] But, given the known difficulties of linking the presence of specific cortical structures with behavior, this remains nothing more than speculation, not direct evidence.

When trying to figure out how to approach the shaky empirical evidence for complex emotions like empathy in dolphins, which is almost exclusively anecdotal in nature, it's possible to apply the Bekoff/Balcombe principle, which suggests that, all things being equal, we're better off assuming dolphins do possess complex emotions like empathy. This is a position that cognitive psychologist Diana Reiss characterizes as a "logical leap" that is "based on a degree of cognitive continuity"[220] between human and dolphin minds, and is strongly argued to be the best approach to the problem of empathy by Frans de Waal.[221] In my opinion, this philosophical approach works when applied to the problem of whether animals subjectively experience basic emotions (insofar as the science is entirely ambiguous on this point), but stumbles somewhat when we're dealing with complex emotions. We suspect that an emotion like true empathy requires something akin to theory of mind. But the evidence from the laboratory does not suggest that dolphins possess theory of mind, and thus logic might work against us in suggesting that they then lack the necessary cognitive machinery to generate empathy. So in this case, all things are *not* equal insofar as there might well be a break in the continuity between human and animal minds on this point. But I want to be clear that this evidence is, at best, inconclusive. In fact, just like the problem of subjective experience, it might be methodologically impossible to ever provide enough evidence to be sure of the existence or non-existence of theory of mind in dolphins. And, while it is likely, it is by no means an established fact that full-blown theory of mind is required to produce something akin to human-like empathy in the dolphin mind in the first place. Perhaps species that lack full-blown theory of mind but have a sense of self and can make a self-other distinction are able to experience some degree of empathy. But quantifying "degrees" of empathy or "levels" of self-awareness in animals is, at present, more of an exercise in speculation than empiricism.

Therefore, a strong claim that dolphins either do or do not possess complex emotions like empathy based on empirical evidence is, in both cases, likely to be wrong. The reality is that we simply don't know. When faced with a formidable epistemological problem like this, the suggestion that science has anything tangible to say at present about whether dolphins possess complex emotions is more a leap of faith than a leap of logic. For better or worse, it's our gut that needs to lead the way on this one.

4 The Proof of the Pudding is in the Behaving

We suck at being able to validly measure intelligence in humans. We are even worse when we try to compare species.[1] *Stan Kuczaj*

Symbol Use

Because humans are expert symbol manipulators, our understanding of intelligence in animals often involves testing for their ability to use language-like artificial symbols systems, if not outright language. After a handful of unsuccessful attempts to teach apes to speak from the 1930s,[2] Allen and Beatrix Gardner published a groundbreaking paper in *Science* detailing the success they had in teaching the chimpanzee Washoe to use gestures (not vocalizations) to communicate.[3] Soon after, similar programs were established for most of the great ape species (orangutans, gorillas, chimpanzees). A controversy soon arose concerning the methods used in these studies, and whether or not the researchers had inadvertently cued their animal subjects to produce signs/sentences, or if the animals truly did intend to produce—and understood—the symbols they were using. It was amid this controversy that Louis Herman from the University of Hawaii began to study the ability of dolphins to comprehend—but not produce—artificial symbols, a method that allowed for tight control of the experimental setup, and results that would be free from the kinds of Clever Hans criticisms that had crippled other animal language work. The subsequent results of Herman's work are likely the best-known examples of complex dolphin behavior driven by complex cognition, and good reason for staking the claim that dolphins are intelligent animals.

A symbol is a kind of stimulus (e.g., visual, auditory) that represents or stands in for an idea or concept, either concrete or abstract. The symbol itself is completely arbitrary, bearing no physical or other resemblance to the thing it represents (unlike an icon, which does resemble what it represents). The human ability to manipulate symbols—including words or numbers—is considered the cornerstone of our intelligence.

Herman's most famous dolphin prodigies, Akeakamai (Ake) and Phoenix, were taught acoustic (computer-generated whistles) and visual (human produced gestures) symbols that represented a variety of objects (e.g., hoop, ball) and actions (e.g., touch, jump),[4] as well as locations, agents, and relationships.[5] A handful of symbols also represented more abstract concepts like *left* and *right*, or the *question* symbol, which indicated that the trainer was asking the dolphin a question that required a yes or no response. Ake was taught the gestural symbol system and Phoenix the acoustic system. The two systems differed in the meaning attributed to the order in which the symbols were produced. The phrase *hoop fetch ball* meant "take the hoop to the ball" in Phoenix's system. In Ake's system, this concept would need to be rendered as *ball hoop fetch*. These systems allowed Herman to test if the dolphins had a mental representation of the objects and actions that the symbols represented, or were simply learning to associate the symbols with specific behaviors. Mental representations are the meat and potatoes of complex thought—a way for the mind to classify entities in the world, attribute meaning to them, and manipulate them in order to solve problems. They form the basis of the hypothetical "language of thought" proposed by Jerry Fodor,[6] which Steven Pinker described as "mentalese."[7]

The outcome of Herman's many years of symbol work produced an impressive laundry list of cognitive skills:

1. Dolphins could follow commands to carry out actions with objects represented as either a symbol, or a replica of the object.[8]
2. The dolphins could follow commands even when the objects were not present, and only given to them after a delay.[9]
3. The dolphins could report the presence or absence of objects in their pool by pressing a yes or no paddle.[10]
4. The dolphins were required to form relationships between multiple objects in a sequence, including the use of modifiers (e.g., left, right) for sentences containing up to five words. For example, *right water left basket fetch* means take the basket on the left to the stream of water on the right.[11]
5. The dolphins needed to wait for the entire sequence to be presented before drawing a conclusion as to the nature of the relationships between concepts, as opposed to responding to each gesture in sequence.[12]
6. The symbolic mental representation of the object could be involved in any number of roles. For example, the concept of *person* could be acted on directly (e.g., leap over the person), indirectly (e.g., bring the surfboard to the person), or simply remain a concept that is not involved in an action (e.g., report on whether a person is in the dolphins' pool or not).[13]
7. The dolphins could respond correctly to novel sentences that used familiar words given in a never before seen sequence.[14]
8. The dolphins understood some symbols as representing a class of object or action—for example, the symbol *ball* would refer to all balls, not just one specific ball.[15]

9. When given symbol sequences that violated the syntax rules of the systems they were trained on, the dolphins either rejected the sentence outright (ignored it), or extracted a meaningful phrase by ignoring certain elements while retaining the semantic relationship for word order.[16]

These discoveries constitute fairly convincing evidence that Ake and Phoenix did have mental representations of the objects, actions, and concepts referred to in their symbol systems. Their ability to extract meaning from word order (a rudimentary form of syntax) is also a unique skill. So how complex and unique are these mental representation skills really? Do they constitute hard evidence of an uncanny intelligence bubbling away in the dolphin brain?

It's possible to argue that the idea of a "mental representation" is not all that uncommon in the animal kingdom, and thus nothing to get all that excited about. In order to survive, all animals must sense aspects of their environment, extract those features that are relevant to their needs, commit them to short- or long-term memory, categorize, and compare different kinds of *things* that they encounter, and make a decision on how to behave based on changes to incoming stimuli. Thus, any time an animal makes a decision that is not triggered by trial and error, they are using some form of a mental representation. It has long been argued that insects likely use mental representations to create cognitive maps used to navigate their surroundings and envision shortcuts, which rely on their brains' ability to encode representations of landmarks, distance, angles, velocity, etc.,[17] although a debate rages in the literature as to how map-like these maps really are.[18] It's not particularly easy to sniff out the presence of mental representation in animal minds, which is why scientists continue to debate whether or not it has been conclusively shown that, as some have argued, pigeons have a concept of a "human being,"[19] chimpanzees have concepts for "food" and "tool,"[20] or bees have a concept of "above" and "below."[21] Nonetheless, it is likely that many—if not all—species could be understood as possessing mental representations of the world around them in some form.

But where symbol use is concerned, Ake and Phoenix were able to broaden the role of the mental representation to include both the real-world concept to which it referred, and to a learned symbol. This *is* something worth getting excited about. Arguably, this ability allows for a more complex means of organizing the language of thought, and is certainly an ability that is less common in the animal kingdom. Species that are able to comprehend multiple symbols to a similar extent include chimpanzees, bonobos, gorillas, orangutans, sea lions, and parrots. Of course, some species, like pigeons and monkeys, can be taught to comprehend symbols that stand in for abstract concepts like "same" and "different," with some controversy as to the extent to which they form and link abstract mental representations of these concepts with the symbols in question.[22,23] In contrast, much of Herman's evidence for symbol use in dolphins, including the ability to reference absent objects, strongly suggests that Ake and Phoenix truly did understand the link between the symbol and its referent, and thus formed a mental representation that linked the two. Some have speculated[24] that this is part of a body of evidence that dolphins can form and think not just about mental representations, but about representations of representations (i.e., metarepresentations or secondary representations).[25] Whether or not you should consider the ability to use symbols in this manner as a sign of dolphin intelligence (i.e., a measure of how closely a dolphin's behavior resembles the behavior of an adult human) is entirely a personal matter, but it is probably a rare ability in the animal kingdom.

Herman also tested the dolphins' ability to respond to images on a television screen.[26] This is a skill that home-reared chimpanzees, including Sue Savage-Rumbaugh's language-trained prodigies Sherman and Austin, reportedly had a difficult time learning in early experiments.[27] Only after intense training did they show any interest in looking at a television screen, much less respond to the images as representations of reality. When researchers placed a small black and white television in a viewing window of the dolphins' tank, everyone expected that Ake and Phoenix would require the same level of intense training that great apes require in order to get them to respond to the screen. But to Herman's astonishment, both Ake and Phoenix responded immediately to the television screens.[28] Ake was able to follow the gestural commands presented on the screen by the trainers without any problem—even when the images were degraded/distorted. Herman's conclusion here was that Ake "had developed rich representations of the gestural symbols and that she used those representations to make sense of stimuli that bore little physical similarity, other than movement pattern, to the gestures displayed by the full-body image of the trainer."[29] Of course, it's difficult to say for certain the extent to which the (degraded) gestural symbols on the video screen were unlike the live version (and thus how generalized or rich the mental representation was)—it might well be that Ake simply understood the trainer as being stuck behind a window,[30] and that the iconic representation from the television screen was not an entirely unexpected or difficult to interpret stimulus for Ake. In fact, stimuli presented via video screens are now widely used to study animal behavior and cognition, with chimpanzees,[31] chickens,[32] and jumping spiders[33] responding appropriately to video images displaying real-world stimuli. Nonetheless, for whatever reason, dolphins appear to be faster/better than great apes at figuring out that television screens can transmit visual representations of reality. Although to be fair, jumping spiders, with their eight eyes and brains so big for their body size that they spill over into their legs, seem equally as skilled.

Concept Formation

Herman's language work revealed that dolphins likely form mental representations of a number of concepts (e.g., presence/absence of objects). Studying concept formation in animals has a long history in the field of comparative psychology, which touches on issues of learning, memory, discrimination, perception, etc. It is a perpetually contentious field of inquiry in that it can be difficult to differentiate between an animal that is forming an abstract, high-level mental representation of a concept, or low-level stimulus generalization. But where dolphins are concerned, many scientists have argued that experimental evidence strongly indicates that they hold higher-level generalizations.[34] In addition to Herman's language work, evidence exists that dolphins are able to comprehend a number of both concrete and abstract concepts, including:

1. The ability to classify objects as *same* or *different* based on their 2D or 3D shape.[35]
2. The ability to classify a relative number of items as being *less/fewer* (e.g., three dots are less than seven dots).[36,37]
3. The ability to classify objects as *larger* or *smaller* than other objects.[38]
4. The ability to create a generalized concept of *person/human* that applies to all people/humans.[39]
5. The ability to classify a series of tones as either descending or ascending in frequency.[40]
6. The ability to discriminate and classify both natural dolphin whistles and artificial whistles, which would allow for the recognition of an individual dolphin's signature whistle.[41,42]
7. The ability to understand that an object that is hidden out of view continues to exist (i.e., the concept of object permanence).[43]

Dolphins also perform well on match-to-sample tasks, which require them to observe an object (e.g., a Frisbee) and then choose the matching object from two or more comparison objects (e.g., a hoop and a Frisbee).[44] They are able to do this when presented with the objects visually (in air or on a television screen) or in a situation where they were required to echolocate on the objects (i.e., the objects were placed behind a visually opaque but echoically transparent barrier). Importantly, they can even manage this task when shown the object in one modality (e.g., visually), and then asked to choose the same object from a group of objects presented underwater that they could only perceive via echolocation, and vice versa. In the many incarnations of this experiment, dolphins do not seem to have any trouble whatsoever with this type of cross-modal object matching.[45,46,47,48,49] This suggests that dolphins possess a mental representation of objects that is not just a list of object features, but a global, integrated representation that transfers easily between different modalities (i.e., across vision and echolocation). Results like these have led some to speculate that the fact that the auditory and visual projection zones in the dolphin cortex are adjacent to each other presumably correlates with their skill on these tests,[50] and could mean that dolphins process information in a more "integrated" manner than primates (whose brains do not have the same level of cortical adjacency), and that they receive and process more sensory information than humans do.[51]

So just how unique or complex should we consider this laundry list of concept formation skills for dolphins to be? Not as unique or complex as one might think. The idea that dolphins' cross-modal matching skill is a sign of a special, integrated information processing capability is probably not correct. The modern science of the brain suggests that cross-modal interactions are the norm when it comes to how animal brains (including human brains) process incoming stimuli.[52] Aside from the fact that many species perform well on studies of cross-modal matching (e.g., primates,[53] dogs,[54] rats[55]), evidence suggests that there is constant interaction and integration of sensory signals throughout the brain, with multisensory convergence that results in an "integrated" processing of information being the rule (and not the exception), and likely unrelated to proximity of cortical projection zones. Consider the infamous McGurk effect. If you watch a video of someone saying the sound /ga/, but accompanied by the audio of someone saying the sound /ba/, your brain will perceive the sounds as

/da/. This is because both vision (looking at mouth/lip movements) and audition are used by your brain to create the perception of a sound (i.e., integrated processing). The McGurk effect cannot be explained by something akin to cortical adjacency causing deeper "integration" of sensory information since the visual and auditory projection zones are not even remotely close to each other in the human sensory cortex (located in the occipital and the temporal lobes respectively). Functional magnetic resonance imaging scans of the human brain when it produces the McGurk effect show a series of brain activity involving a number of non-adjacent brain areas, including the primary and non-primary audio cortices, the visual cortices, the superior temporal and intraparietal sulcus, etc.[56]

And consider also the neurological condition synesthesia, in which input from one sensory pathway triggers the production of activity in a seemingly unrelated pathway. This leads to a synesthete, for example, tasting eggs every time he/she hears the phoneme /k/,[57] or seeing the color orange when hearing music in the key of D major. These sensory integrations occur between senses with projection zones across the cortex, not just those zones that are near each other, which suggests that sensory integration for a synesthete is underpinned by something other than cortical proximity. Even for us non-synesthetes, integrating sensory information and recognizing/matching an object across senses with distant cortical projection zones, like touch and vision, is fairly straightforward.[58] Given all this, it would have been rather surprising if dolphins—with known high acuity for both their visual and echolocation sensory systems—did not perform well on tasks of cross-modal matching. There is no need to attribute extraordinary brain sophistication to these results, although they do confirm the rather wondrous sensitivity of echolocation, and do tell us more about how animal brains produce mental representations of object/concepts based on sensory input and object features.

It turns out that most of these concept formation skills found in dolphins should perhaps not be considered all that exceptional. Like dolphins, many species perform well on tests of relative quantity judgment (i.e., the ability to classify groups of items as being less/fewer), including sea lions,[59] elephants,[60] black bears,[61] squirrel monkeys,[62] and mosquitofish.[63] Salamanders are able to differentiate between small quantities of items (as long as they are fewer than four),[64] and honeybees keep a running count of landmarks that they fly over in order to help them navigate.[65] Like dolphins, pigeons can easily learn the concept of *same/different,*[66,67] and learn to discriminate and categorize object concepts like *chair, human, car,* and *flower.*[68] Pigeons can even learn to discriminate between paintings by Van Gogh and Chagall with the same accuracy as college undergrads.[69] In a rather peculiar experiment, pigeons were taught to discriminate between good and bad art, and thus might have learned the concept of "beauty."[70] All of these studies suggest that many animals, not just dolphins, can be taught to classify and categorize the world in surprisingly human-like ways, and that basic numerosity (e.g., small and relative quantity judgments) is possibly an evolutionarily ancient skill that is found across the taxa.

It's also worth noting that dolphins' understanding of object permanence (i.e., the idea that objects continue to exist even when they are not directly observable), is not quite as advanced as one would expect. Given their abilities to form mental representations about objects as witnessed in Herman's symbol experiments, it was expected that a dolphin would be able to pass Stage 6 of Piaget's object permanence scale: being able to track the position of an object that was hidden and then moved to another location (i.e., invisible displacement). In the one published report of dolphins' attempts to solve this problem, they quite clearly failed.[71] They had no trouble keeping track of an object that was moved and then hidden under a container (i.e., visible displacement), but failed to keep tabs on the object if that container was then itself moved. This is a skill that other species, including chimpanzees, bonobos, gorillas, orangutans,[72] dogs,[73] and possibly even magpies, parrots, macaws, parakeets, and cockatiels[74] possess. By all accounts, dolphins should have solved this task. Whether they failed because of the nature of the experimental design, or because they really do lack the ability to track hidden objects (perhaps because of their reliance on echolocation to perceive hidden objects), remains to be seen.

In summary, it's simply not the case that dolphins outclass all other species when it comes to the conceptual skills listed here. In some tests, pigeons and parakeets have bested dolphins. But dolphins are ahead of the pack when it comes to conceptual skills as it pertains to symbol use, which suggests that the richness and complexity of the concepts and mental representations present in the dolphin mind might be of a different nature than that found in, for example, the pigeon mind, and more on par with that of primates and parrots. This, however, is a difficult thing to judge, and it is likely that future experimental results of conceptualization skills for dolphins and other species will continue to alter our view of how animal minds make sense of the world around them.

Memory, Planning, and Innovative Problem Solving

Working memory (i.e., short-term memory) is closely linked with performance on numerous cognitive tasks for humans and animals, and is often directly linked to the concept of intelligence. Problem solving, reasoning, etc. require the brain to access and manipulate information stored in working memory. It might be possible in some animals[75] (but perhaps not humans)[76] to improve performance on cognitive tests by improving the efficiency of working memory. Studies of working memory in dolphins reveal that they can, while working within the match-to-sample task described earlier, remember an object presented visually for up to eighty seconds, and remember a sound for up to three minutes.[77] They can also remember a list of at least four items,[78] which is just shy of the seven plus or minus two standard for humans tasked with remembering serial lists, and similar to the number for primates and pigeons.[79] What does this say about dolphin intelligence? Not much. While some have labeled dolphin memory as "impressive"[80] based on these experiments, there is no comprehensive comparative study of working memory in animals to indicate whether dolphins might have a superior or inferior ability to hold information in working memory.

Studies of long-term memory in dolphins don't give us much more to chew on. Dolphins are obviously capable of remembering learned behaviors for a lifetime, and can recall arbitrary associations between symbols and their referents (e.g., Herman's gesture language) over the course of many years. But there are no direct tests of dolphin long-term memory to see either how long or how many associations they can remember. In one such study for other species, it was shown that pigeons and baboons could learn to recall up to 1,200 and 5,000 (respectively) individual pictures together with their "appropriate" response (i.e., touch the target on the left or the right of the screen) for up to five years.[81] Would dolphins perform as well as the pigeons and baboons on these experiments? It remains to be seen.

In terms of complex—and possibly uniquely human—kinds of memory, episodic memory is the one to beat. Episodic memory is defined as "the recall of a specific event in the past, which the rememberer has the sense of having personally experienced."[82] Going beyond simply recalling semantic knowledge (e.g., facts) about the past, episodic memory encodes a large amount of personal information about particularly salient events. For example, knowing that dogs have sharp teeth is semantic knowledge, but recalling that awful time when your neighbor's dog bit your hand when you tried to take away his squeaky toy involves episodic memory. At the moment, there is debate in the literature as to whether or not any animals other than humans have true episodic memory, with the phrase "episodic-like" used to describe instances in which animals display behavior that could (but might not) be a result of episodic memory recall. It's not well understood the extent to which an animal would need to be self-aware, hold metarepresentations, have a theory of mind, or have language abilities in order to experience episodic memory in the same manner as a human. Endel Tulving, who first proposed the distinction between semantic and episodic memory, hypothesized that some form of self-awareness, which he termed autonoetic consciousness, was necessary for episodic memory.[83] It may be that animals are able to remember the facts about past events (i.e., semantic memory), but are unable to view it as a "personal experience," or the "personal past," which is similar to how very young children (i.e., under the age of three or four) understand past events. The development of language skills and theory of mind in children coincides with the development of episodic memory, which makes this an attractive hypothesis. Since testing for the presence of phenomena like self-awareness, theory of mind, and other subjective experiences in species that lack the language skills to express their thoughts is, as we have seen, extremely challenging, it should come as no surprise that scientists have yet to uncover unequivocal evidence of episodic memory in animals.

A handful of experiments have provided results that hint at episodic-like memory in dolphins. When given a command to *repeat* a complex behavior that they've just performed (e.g., swim in a circle with fins waving) the Kewalo Basin dolphins could do so easily.[84] In addition, the dolphins were first asked to perform a novel behavior (i.e., one that had not recently been performed), and then asked to repeat that behavior. In doing so, the dolphin needed to recall detailed aspects of the behavior itself (i.e., answer the question "what was the relatively novel response that you just made?"),[85] and not just recall the gestural command associated with that behavior. They performed equally as well in another experiment where they had to repeat their actions in combination with objects and locations.[86] These results suggested that the Kewalo Basin dolphins must have had fairly complex mental representations of recent past events, and that, to some extent, they might have placed themselves at the center of the memory insofar as they were able to self-imitate their own behavior.

But the strongest evidence of episodic-like memory in animals comes not from dolphins or the great apes, but from scrub jays.[87] Scrub jays, like many birds, cache (i.e., hide) food items that they later retrieve when hungry. The jays understand that some foods decay quickly (e.g., worms), while others can be stored for long periods of time (e.g., nuts). If, after caching these foods, enough time has gone by that their favorite food has likely spoiled (a concept they are capable of learning), they will not bother to visit the areas in which those foods have been cached, which suggests that they have some representation of both the passage of time, and approximately when they cached the food in the past. Scrub jays appear to remember which food types they have recovered from cache sites, and change their recovery plan depending on what they find at the cache sites (e.g., finding rotten worms means stop visiting cache sites with worms). The jays appear to remember what they have cached, where they have cached it, when it was cached, and understand that the "what" and the "where" can remain the same, but the "when" can be different. They might even retain information about "who" was watching them make their food caches, moving their cached food if a dominant scrub jay (as opposed to a subordinate) had been watching them at the time of the initial cache.[88] Rhesus monkeys, rats, gorillas, squirrel monkeys, mice, chimpanzees, and meadow voles have all displayed somewhat similar abilities, although the evidence from scrub jays is often considered the strongest case for episodic-like memory in animals.[89,90,91]

The memory of past experiences—whether stored in short-term memory, long-term memory, or taking the form of episodic memory—is central to how animals cope with present and future events. Using past experience to guide present and future behavior is the foundation of the many forms of learning that animals exhibit. Within this domain, the most sophisticated and intelligent (i.e., human-like) use of memories is to solve current and future problems. Any major achievements in human history revolved around our ability to provide a novel solution to a (sometimes novel) problem, whether it's how to find food or how to send people to the moon. Of course, all animal species solve the problem of how to find food, but very few of them rely on episodic memory (like the scrub jays) to do it, instead basing their actions on instinct, trial and error, or associative learning. Humans avail of these skills too, but can also lean on a set of complex cognitive skills that include:

1. mental time travel (including anticipation of future mental states)
2. planning
3. insight.

In theory, if an animal (like a human) has the ability to form episodic memories and thus remember its personal experience of the past, it should also be able to project itself into the future and envision possible future events (i.e., mental time travel). If the animal also has a concept of self and is able to form metarepresentations, including being aware of its own mental states (which may or may not be a prerequisite for mental time travel), it might then be able to anticipate future mental states (like hunger) that it is not currently feeling. Combining these abilities will allow the animal to solve anticipated problems by "envisioning" a solution in the mind's eye (i.e., insight) by thinking about both past personal experience (i.e., episodic memory, mental time travel) and possible future events and mental states (i.e., mental time travel). Humans are clearly capable of all of these things. Animals on the other hand might display certain combinations, but have yet to unequivocally prove that they combine all of these skills.

For example, a great many behaviors observed in animals could be described as involving "planning" for future events, including squirrels storing nuts for the winter or bears hibernating. However, almost all of these behaviors are instinctual, inflexible behaviors that do not require the individual animal involved to have any concept of past or future events (e.g., a squirrel will store nuts for the winter even if it has never experienced winter beforehand), nor any concept of its potential future mental states. And insight problem solving might well occur in animals (e.g., Wolfgang Köhler's famous experiment in which chimpanzees suddenly realized that they can fasten two sticks together to reach a banana that is out of reach[92]), but these are often within the context of solving an immediate (as opposed to future) problem (i.e., I am hungry NOW, so how do I get that banana). Within experimental contexts, it is possible to get animals to display behaviors that look like insightful planning that involved mental time travel, but that can often be a result of an experimental design that (accidentally) reinforced certain learned behaviors or associations. For it to be true planning, the behavior must be based on the memory of a past experience, and be used to address a future (as opposed to immediate) need. For example, a cat might remember the time when its owner left for the weekend and forgot to leave enough water in its bowl. To make sure it doesn't go thirsty again, the cat could demonstrate true planning by storing full bowls of water under the bed (even though the cat was not thirsty at the moment) that she could drink from in the future in the event that the careless owner should forget to leave her enough water.

A possible example of true planning is the now infamous story of Santino the male chimpanzee at Furuvik Zoo, who diligently stockpiled stones and other projectiles in the morning before the zoo opened so that he could hurl them at unsuspecting zoo visitors later in the day.[93] In recent years, Santino has taken his behavior one step further. Since visitors now know to be on the lookout for stones hurtling at their heads, Santino has taken to hiding the stones from visitors' sight behind logs or hay that he collects from his enclosure.[94] He makes sure to hide the stones when visitors are not around, and then sits near the logs/hay pile waiting for them to get close enough, grabbing for the stones in a surprise attack. Although this is just one series of observations on a single animal, it suggests that Santino was improvising novel solutions to a future problem, and may have been aware of the attentional states (and maybe even the minds) of the visitors at the zoo whom he was plotting to deceive and ambush. He may have been anticipating his own future mental state (i.e., angry enough to want to throw stones), and projecting himself into a future situation that he had not yet experienced (i.e., an ambush scenario) based on his experience of similar, but not identical, past situations.

Much experimental work has been conducted in the past couple decades to determine the extent to which animals might exhibit mental time travel, planning, and insight, with recent evidence suggesting that these abilities might well be present individually (if not altogether) in some species.[95] Controlled experiments have shown that bonobos, orangutans, and chimpanzees collected, saved, and transported tools that they were going to need for food collection in the future—including preparing them the night before.[96,97,98] The apes involved in these experiments were also clearly planning for a time in the near future when they would find themselves in a similar situation to one they had experienced in the past that required the use of tools to help solve their hunger issues. Other species, like squirrel monkeys,[99] provide even more convincing experimental evidence that animals can anticipate future events. In these experiments, the monkey had just drunk its fill of water and was made to choose between a small or large bowl of food—either of which would make them thirsty again soon after eating it. If the monkey chose the large bowl (which is something they would naturally want to do), it would not receive water again for a while. But if it chose the smaller food bowl, the water would be returned quickly. They soon learned to choose the smaller bowl, which likely meant that they were anticipating being thirsty in the near future, and wanted to make sure they had water on hand. This might well be evidence that the monkeys were anticipating a future event—and their own future mental state.

But it's scrub jays again that provide the best evidence from the lab that mental time travel is involved in future planning for some animal species. In one experiment,[100] jays were presented with a bowl of cacheable food in the evening. In the six days prior to this, they had been kept in one of two nearby compartments overnight: one in which they had received breakfast in the morning and one where they had not. Presumably the jays anticipated that, if kept in the no-breakfast compartment that night, they would end up hungry in the morning (i.e., episodic memory of past events combined with future projection and mental state anticipation). So as a solution (i.e., insight problem solving and planning), they cached three times as many nuts in the no-breakfast compartment as in the compartment where they had previously received breakfast.

For dolphins, much of the evidence of problem solving and planning comes from anecdotal or observation accounts as opposed to experiments.[101] Often, these observations, like that of Santino the chimpanzee stockpiling projectiles, make it difficult to determine if the animal has truly engaged in mental time travel based on rich mental representations involving the projection of oneself or one's behavior into the future (as is suspected for the scrub jay experiment), or if the animal achieved planning success due to simpler cognitive processes or events, like dumb luck, serendipity, or trial and error learning. Before we can be certain that an animal's behavior is the result of planning and insight, one needs to be 100 percent sure of how and when the behavior came about. Outside of the controls of a laboratory setting, it's difficult to know for sure just how an animal might have learned a behavior.[102] Possible examples of problem solving and planning in dolphins include male dolphins cooperating to herd female dolphins for easier mating access,[103] killer whales working together to wash seals and penguins off of floating ice,[104] and dolphins using sea sponges on their rostrums to help them forage for food.[105] Bottlenose dolphins use air bubbles to confuse and capture prey,[106] and will strand themselves on muddy sloped beaches in cooperative hunting groups in order to capture fish.[107] Some captive killer whales have been observed luring seagulls into their pool by using pieces of fish to attract them, then grabbing the gulls.[108] One captive rough-toothed dolphin used an object (i.e., a swim fin) to prop open a gate between two adjacent pools, and another used a buoy as an underwater projectile to dislodge a walkway.[109] A single wild Indo-Pacific bottlenose dolphin in Australia has been observed hunting for cuttlefish using a peculiar—and perhaps unique—hunting technique that is likely her own devising. After first killing the cuttlefish, she beat the ink out of it, scraped it in the sand to remove the skin and the cuttlebone, and consumed the rest.[110] This complex series of novel behaviors might well have been a result of insight and planning on the dolphin's part, although trial and error and serendipity cannot be ruled out. Unfortunately, none of these examples can be conclusive evidence of insight and planning insofar as researchers were simply not there to observe how the animals first hit upon the idea of the behavior in question.

However, there is evidence of problem solving from the lab where the origins of planning behavior are well documented. For example, Ake's response to anomalous gestural sequences could be considered problem solving insofar as she had to devise a response strategy based on her knowledge of the syntax system of her artificial gestural language.[111] And when Ake was given the command to *fetch* the many objects floating in her pool after a long day of training, Ake understood that this referred to all objects in the pool, and executed a retrieval pattern that saw her rounding up several objects at once, starting with those objects at the back of her pool.[112] This is a solution she appears to have come up with all on her own. Ake also learned to produce a distinctive loud whistle to alert nearby staff that she had found a piece of debris in her pool, and held it in her mouth for the staff member to retrieve. This behavior was rewarded with a fish, but had never been explicitly taught, and suggests that Ake had come up with innovative solutions to problems she faced in her environment. But, like all the other observations, it's simply impossible to know the extent to which mental time travel, planning, and insight were involved.

One of the few controlled experiments to test dolphin problem-solving abilities involved two dolphins, Bob and Toby, housed at Disney's The Living Seas.[113,114] In the first task, Bob and Toby were presented with a situation in which they had to retrieve four weights and drop them into a box in order to release a fish reward. They had watched a human diver perform the task previously by picking up one weight at a time and bringing it to the box, with the fish only being released after all four weights were in the box. The dolphins quickly learned to perform the same behavior, placing each weight in turn into the box. But when the weights were placed far away from the box (i.e., forty-five meters), both Bob and Toby devised a solution to this task that would result in them getting the fish reward faster: Toby collected two weights at a time, and Bob collected all four weights together and dropped them together into the box. This technique had never been demonstrated to them before, so appears to have been the result of a flash of insight. In the second task, the dolphins were given a single weight that they needed to reuse in order to obtain a fish reward by dropping it in three separate boxes. However, two of the boxes had open bottoms, and one of them had a closed bottom, which meant that the dolphin would be unable to retrieve the weight and bring it to the next box(es) if it deposited the weight in the closed bottom box first or second. Both Bob and Toby quickly realized that dropping the weight in the closed-bottom box last was the only way to be sure to get all possible fish rewards—a conclusion that was not a product of trial and error, but of insight and planning due to the fact they almost immediately hit upon this solution. It has also been argued that the spontaneous body pointing behavior produced by Bob and Toby (described earlier) arose as an insightful solution to the problem of how to indicate the location of a food reward to human divers. However, in a third problem-solving test, in which the dolphins needed to insert a stick into an opening that was exposed within fifteen seconds of dropping a weight in the box, they were unable to hit upon the expected solution of first bringing the stick closer to the box (making it easier to retrieve it in time) and then dropping the weight.

The successful results of the first two weight/box experiments suggested to the researchers that "the dolphins that we studied were able to create successful plans when faced with tasks unlike any that they would normally encounter in the wild. This suggests a fairly generalized and flexible ability to plan, which in turn suggests that dolphin evolution resulted in a set of cognitive skills that extend beyond those needed to deal with specific problems that dolphins encounter in everyday life, such as mating, foraging and the rearing of young."[115] Insight problem solving similar to what Bob and Toby displayed has been found in many species, including Asian elephants,[116] chimpanzees,[117] keas,[118] New Caledonian crows,[119] and rooks.[120] But unlike the scrub jay or squirrel monkey experiments, Bob, Toby, and these other species had not been making a plan that involved them anticipating a future mental state (i.e., they were solving an immediate problem for an immediate need). And the extent to which the insightful problem solving displayed by, Bob, Toby, Santino, scrub jays, or any of the other species mentioned so far required human-like skills involving episodic memory, mental time travel, metarepresentations, autonoetic consciousness, self-awareness, metacognition, etc. is still an open question. As it stands, the evidence for dolphin problem solving, unlike the evidence from the great apes and corvids, does not suggest that they are envisioning multiple scenarios in their minds as to how to solve a problem and then picking the one that is most likely to achieve the desired outcome. It's of course entirely possible, if not probable, that they do, but the laboratory evidence is too limited to allow us to conclude that dolphins are unique or superior when it comes to solving problems or planning for the future insofar as many species might well be displaying similarly (or more) complex cognition.

Play

A knack for solving problems—like the kind Bob and Toby faced with the weights and the boxes—might predispose an animal to seek out and engage in play. Play might be a way for dolphins to practice problem-solving skills or otherwise keep their minds active. But what is play exactly? Play, like both intelligence and pornography, falls under Justice Potter's maxim "I know it when I see it." It is notoriously difficult to differentiate play behavior from identical behaviors observed during socializing, mating, foraging, etc., which has led some frustrated scholars to declare that play is impossible to study scientifically, or, with a wholly defeatist attitude, that it doesn't even exist. Of course, most scientists (or anyone who enjoys a bit of common sense) will agree with Gordon Burghardt's assertion that "it exists. Accept it and move on."[121] Burghardt, an expert on animal play, has suggested that play can be characterized by five criteria,[122] which can be neatly distilled into the following definition: "Play is repeated behavior that is incompletely functional in the context or at the age in which it is performed and is initiated voluntarily when the animal or person is in a relaxed or low-stress setting."[123] However much I might like the sound of this definition, psychologist Robert Mitchell is probably still (and forever will be) correct in stating that "play is the hobgoblin of animal behavior, mischievously tempting us to succeed in what, judging from the number of failed attempts, seems a futile task; defining play."[124]

The function of play is equally as perplexing as its definition, and might involve gaining knowledge of objects or the environment, practicing behaviors for later use, establishing social connections or social hierarchy, honing motor skills, a means for testing flexible problem-solving skills in a safe context, and as preparation for dealing with unexpected situations. Of course, it might be that play's only immediate function is to have a bit of "fun," but good luck defining "fun" in a scientifically useful way. In any event, "fun" play behavior costs time and energy, and might even put an animal at risk of injury or death, so the rewards of the "fun" must outweigh the risks, which still demands an adaptive explanation. Understanding the function of play as it pertains to its evolution is as sticky a topic now as it ever has been, but for a discussion of dolphin intelligence, the link between play and problem-solving skills is both relevant and important, regardless of its biological function.

Dolphins are considered to be extremely playful animals. The internet is awash with video evidence to support this idea, showing dolphins playing with bubbles, balls, hoops, cats, dogs, whales, seaweed, sharks, plastic bags, humans, boats, and even iPads. From the scientific literature on the subject of play in captive and wild dolphins, we find evidence of locomotor play, involving surfing, aerial displays, erratic swimming,[125] social play including fighting (pushing, biting, ramming), chasing, and socio-sexual play (i.e., usually involving a penis). The literature describes abundant evidence of object play involving harassing other species (e.g., other dolphins, seals and sea lions, sea turtles, fish, etc.),[126] as well as playing with seaweed, sticks, and other floating objects. Dolphins play fairly advanced games including "seaweed keep away" and "ball toss" with humans and other dolphins,[127,128,129] and have been observed plucking feathers from unsuspecting pelicans floating on the water.[130] Two captive rough-toothed dolphins invented a game in which they would take turns towing each other with a hula hoop around their pool.[131]

But the form of dolphin play that is likely the most cognitively complex is referred to as bubble play. Bubble play is not all that common, and has only been observed in captive Amazon river dolphins, beluga whales, and bottlenose dolphins. These species produce bubbles using a variety of techniques, including smacking the surface of the water with a stick to produce bubble curtains,[132] forming and releasing bubbles with their mouths,[133] or releasing air from their blowholes. They can manipulate the bubbles, turning them into bubble rings, bubble trails, and swirling bubble vortices and spirals. The dolphins sometimes simply observe their bubble creations, but often interact with them. Sometimes they swim through the bubble curtain or large bubble rings, or release smaller bubbles to travel up through the larger bubble rings. And, as can be observed in a popular online video of bottlenose dolphins from SeaWorld Orlando,[134] they will manipulate (e.g., turn, flip) the swirling ring vortices by pushing them through the water with pressure waves, and reshape or destroy them by biting them or touching them with their rostra, flukes, and pectoral fins. They will even inject more air into the ring to increase its size, and attempt to join two bubble rings together.[135] Dolphins will also increase the difficulty of their bubble games, presumably to make them more challenging (i.e., fun/interesting). Dolphins as young as one month and as old as thirty-seven years have been observed engaging in this type of bubble play.[136,137]

These bubble games are entirely of their own devising, and ostensibly require the dolphins to plan their behavior, and likely envision solutions to novel problems (e.g., how to blow small bubbles in just the right way to make them travel up through a bubble ring). Of course, it is difficult to know the extent to which associative learning or observational learning (e.g., a dolphin calf will learn by watching its mother produce bubble rings) underpins these bubble behaviors, as this would suggest less insight and more "low-level" cognitive processes at work. However, one study found that dolphins actively monitor the quality of the bubble rings that they produce, which suggests that they are indeed anticipating and planning their behavior in some detail.[138] In a version of the bubble ring game where a dolphin creates one ring, waits a few seconds, and then produces a second ring that is intended to join with the fist ring, the likelihood of this happening is greatly increased if the quality of the first ring is high (i.e., it's a well-formed ring). As it turns out, the dolphins in this study were far more likely to create a second ring if the first ring was of high quality, which suggested that they were paying close attention to the quality of the rings, and adjusting their own behavior accordingly. Of course, this behavior could also be a result of associative learning devoid of any real insight or planning, but it is not unreasonable (and perhaps parsimonious) to think that a dolphin is monitoring its own behavior in this manner given the experimental evidence that they are capable of doing so.

Play is ubiquitous in the animal kingdom, with many species displaying behaviors that fit Burghardt's definition of play, including primates, canids, ungulates, rodents, pinnipeds, birds, and even reptiles and quite possibly insects.[139,140] So although dolphins do tend to spend a fair amount of their time playing, it is not the case that dolphins are unique when it comes to displaying play behavior. However, complex play in dolphins (like bubble play) and other animals does suggest (although not necessarily prove) the presence of complex cognitive processes related to planning, problem solving, mental time travel, etc.

Tool Use

The satirical newspaper *The Onion* once reported the end of human civilization was upon us after discovering that dolphins had evolved opposable thumbs. Now that dolphins had the necessary digits to make and use tools, they were capable of producing "coral-silicate and kelp-based biomicrocircuitry," and would soon enslave the human race.[141] In reality, an animal does not need thumbs to use tools, and perhaps the most prolific (in terms of number of species) kind of animal producing and using tools are in fact thumbless birds. Non-human primates too are tool users, and so are elephants, rodents, bears, sea otters, squirrels, cephalopods, fish, and even crustaceans and insects.[142,143,144] It might surprise some to learn that insects are tool users, although this has long been known. The oft-quoted idea that tool use was considered unique to humans until Jane Goodall discovered wild chimpanzees fishing for termites with twigs[145] is rather a misrepresentation of history. Goodall's discovery was perhaps the first scientific documentation of tool *manufacturing* by a non-human animal, but scientists going back to at least Charles Darwin (who wrote about monkeys using stone hammers and elephants using twigs to swat flies in *On the Origin of the Species*) have known that animals other than humans are tool users.

There is an ill-defined continuum of tool use complexity stemming from low-level cognitive processes (e.g., Hawaiian boxer crabs brandishing a sea anemone in their claws during interspecific fights, presumably an instinctual behavior)[146] up through complex cognition driving the manufacturing and use of tools to solve complex problems using insight (e.g., New Caledonian crows shaping metal hooks to retrieve food).[147] Tool manufacturing involves an animal actively changing the structure of an object so that it is better adapted to perform the task at hand, and, assuming the manufacturing is not a product of associative learning, might require something akin to insight in order to occur. Depending on your definition of a tool, and how much goal-directed "thinking" goes into manufacturing or using tools, tool use is either rare in the animal kingdom (e.g., only a handful of species actively manufacture tools), or fairly common. Is, for example, a lion using a tree as a tool when it sleeps in the shade it produces? Is a bird's nest actually a tool? Is a dolphin using seaweed as a tool when it plays "seaweed keep away?" According to one regularly cited version of tool use, these are not examples of tool use insofar as tool use is "the external employment of an unattached or manipulable attached environmental object to alter more efficiently the form, position, or condition of another object, another organism, or the user itself when the user holds and directly manipulates the tool during or prior to use and is responsible for the proper and effective orientation of the tool."[148] But even this fairly specific definition easily allows for the conclusion that tool use is fairly widespread in the animal kingdom.

So where do dolphins feature in the tapestry of animal tool use? Dolphins are animals with a seemingly complex understanding of objects, demonstrated by their ability to manipulate, name, categorize, and report the presence/absence of them, as well as use objects in play contexts (e.g., complex bubble ring play). But simply interacting with objects or having complex mental representations of objects, is, following the definition provided, not the same thing as using objects as a tool. A tool needs to be used to do something specific (i.e., "alter more efficiently the form, position, or condition of another object, another organism, or the user itself"), which excludes examples of simply interacting with or playing with objects. Dolphins do, however, use tools in a way that quite clearly fits this definition.

In 1997, scientists studying Indo-Pacific bottlenose dolphins in Shark Bay, Australia published the first account of tool use for a dolphin species.[149] A small number of dolphins (just fifty-five)[150] have been observed carrying marine basket sponges on their rostra (a behavior termed "sponging"), which they use to protect themselves from sharp objects (e.g., animals with venomous spines) as they probe the sandy sea floor in search of barred sandperch hidden in the sand—a species that is difficult to locate with echolocation.[151,152] The dolphins need to drop the sponge to capture the prey that they flush out, picking the sponges up again later to continue foraging. Once the sponge has lost its protective value, they will replace it with a new one that they break off of the seafloor where the sponges grow. This is a behavior that is practiced almost exclusively by female dolphins, and appears to be transmitted vertically from mother to daughter—a result of the young animals learning the technique by observing their mothers.[153,154] Sponging originally appeared to have been "invented" by a single female dolphin relatively recently (approximately 180 years ago),[155] who has been given the moniker "Sponging Eve." However, it seems that sponging was discovered independently by a second Sponging Eve located 100 kilometers away living in a similar ecological situation. Sponging is a time-consuming behavior, occupying 17 percent of the spongers' time, and is observed almost exclusively in those animals foraging in deep water channels, where sponging might constitute a means of exploiting an otherwise unused niche. This behavior clearly fits the definition of tool use in that the dolphins are "altering" the form of their rostra to make them more efficient at foraging. It is not an example of tool manufacturing, however, in that the dolphins did not modify the structure of the sponge to make it a more effective tool.

So is this evidence that dolphins are unique tool users, and that their tool use is indicative of intelligence as some suggest?[156] On one hand, sponging is a complex behavior insofar as it was very likely a learned behavior that one (or two) dolphin(s) devised as a solution to a problem (or possibly stumbled across accidently), which has since been passed down to subsequent generations via learning/teaching/culture. This is a more complicated scenario than, say, Egyptian vultures which use stones to hammer open ostrich eggs, a behavior that appears to crop up instinctively without any flashes of insight or teaching involved.[157] Dolphin tool use might be analogous the use of a stone hammers by bearded capuchin monkeys in Brazil to crack open palm nuts, a technique that is passed down through the generations by observational learning, but the origins of which are unknown.[158] For all we know it was alien visitors who first taught capuchins to smash nuts and dolphins to dig for fish with sponges, so we can't really be certain how much complex problem solving Sponging Eve(s) engaged in. The fact that sponging developed on two separate occasions within a small population suggests that it was in fact a product of problem solving and insight, but this is by no means irrefutable evidence.

Given this scenario, dolphin sponging is not a clear example of an animal manufacturing a tool to solve a novel problem via insight. First, there is no manufacturing involved, and second, we do not know the origins of the behavior. Sponging Eve might have stumbled upon the technique by accident after getting a sponge stuck on her rostrum. Crows, rooks, and chimpanzees, on the other hand, have clearly demonstrated their ability to manufacture complex tools to solve novel experimental problems, which cannot be a result of serendipity or trial and error. Given dolphins' ability to solve problems in the laboratory (e.g., Bob and Toby's weight/box experiment), it should be hypothetically possible (and perhaps even probable) that dolphins could manufacture a tool to solve a similar problem. And it's still arguable that dolphin anatomy is less suited to fine-scale object manipulation than, for example, the beaks and claws of New Caledonian crows, which is why they have not been observed engaging in tool manufacturing. Perhaps with the exception of the very specific habitat at Shark

Bay where sponging occurs, the ecological conditions for most dolphin populations are simply not ripe for the development of tool-using behavior. Under the right conditions and with the right objects at their disposal, they might be master tool producers. But this has yet to be demonstrated. Of course, it's always possible that wild dolphins regularly manufacture and use tools, but that humans have simply not been around to witness it. In any event, it is up the reader to decide whether dolphins should be penalized for this lack of tool-manufacturing evidence when assessing how intelligent (human-like) their tool-using behavior is.

Culture

At a conference focused on the issue of cetacean rights held in Helsinki in 2010, a group of scientists, philosophers, and animal rights activists produced a document titled the Declaration of Rights for Cetaceans.[159] Among the ten conclusions listed in the declaration was that "cetaceans have the right not to be subject to the disruption of their cultures." The idea of "culture" is often thought of as a uniquely human trait, and there is currently a debate in the scientific community as to whether or not animals could truly be described as possessing "culture."[160] This is partly a problem of definition, with ethologists, comparative psychologists, anthropologists, sociologists, biologists, etc., disagreeing on what criteria are required for social behavior to earn the label of "culture." All likely agree that culture must include some form of social learning that allows behaviors to be transmitted between individuals within a population. But nobody can agree on whether this requires teaching, imitation, how many animals in a population (or subpopulation) need to have learned the behavior,[161] and/or if it requires the establishment of something akin to social norms (i.e., informal laws as to how one is expected to behave), ratcheting (i.e., increased complexity/functionality of traditions),[162] or human-like concepts of shared-identity and shared-values. If culture could be described as simply the transmission of acquired information, then bacteria could fit the criteria.[163] If we use an anthropologist's or sociologist's definition, which often includes shared belief, value, and symbol systems, then no non-human animal qualifies.

The idea of culture as it applies to dolphins—and cetaceans in general—burst onto the scene in an article authored by Luke Rendell and Hal Whitehead published in 2001.[164] The definition of culture adopted by this article, was "information or behaviour acquired from conspecifics through some form of social learning."[165] This definition is broad enough to encompass a large number of socially learned behaviors in a number of species. The general idea is that culturally acquired behaviors are not a product of genetics or are something that would spontaneously crop up in a population based solely on specific properties of their environment/ecology that each individual learns to adjust their behavior to independently (i.e., without learning/modeling from conspecifics). Because of the pushback by some disciplines that would like to continue to use the word "culture" only to describe what humans do, some scientists working with dolphins have suggested it might be better to use the seemingly less controversial term "traditions" to describe what happens when information is passed via social learning,[166] although others have argued forcefully that culture is the correct term.[167]

The most famous case of culture in cetaceans is that of humpback whale song.[168] Humpback whale populations from the same geographic area/population sing an identical song on their winter breeding grounds, which will continue to change in form over the course of months and years. Despite these changes, individuals will all continue to sing the same communal song. According to Whitehead, "there is no conceivable mechanism for such patterns other than animals listening to each other's songs and adjusting their own accordingly to produce stereotypical group behavior—culture."[169] As for dolphins, Rendell and Whitehead's 2001 article suggests a number of behaviors that they argue fit their definition of culture, including:

1. Sponging behavior in Shark Bay—a cultural tradition passed down vertically from mother to daughter via social learning.
2. The vocal dialects of killer whales—where matrilineal groups share a set of discrete calls that are likely the result of social learning as opposed to genetics.
3. The variation in (often cooperative) hunting techniques and group-specific hunting grounds seen in both killer whales and bottlenose dolphins around the world.
4. Pod specific "greeting ceremonies" seen in killer whales.
5. The provisioned dolphins at Monkey Mia in Australia, where young dolphins might learn to accept food from humans by watching their mothers.
6. Bottlenose and Irrawaddy dolphins cooperating with human fisherman at Laguna Brazil and in the Ayeyarwady River, Myanmar (respectively).
7. The hunting techniques of killer whales from the Crozet Islands, that teach their calves to intentionally strand on the beach in order to capture seals.

Of all of these examples, the strongest case for culture in dolphins comes from the vocal dialects of killer whales, which are almost certainly a product of social learning. The acquisition of signature whistles by bottlenose dolphin calves is also a clear product of social learning. But it is not the case that all dolphin scientists agree that this short list of dolphin behaviors fits even the broad definition of culture in animals provided by Rendell and Whitehead. By all accounts, dolphins, with their skill at imitation, mimicry, and observational learning as proven in laboratory settings (both in the vocal and motor domains), prolonged interaction between mothers and calves, and complex social structure, should be particularly primed for benefiting from a social learning scenario. But examples of observed wild behavior make it difficult for researchers to differentiate between a socially learned (cultural) behavior and one that is a product of individual learning driven by ecological circumstances or genetics. For example, we do not know if the Monkey Mia dolphins learn to accept food from humans from watching other dolphins, or by having their behavior shaped by the behavior of the humans on the beach—completely independently of what the other dolphins are doing.[170,171] Although a recent study suggests that bottlenose dolphins around Perth, Australia, learn to accept (illegal) handouts from humans via social learning.[172] But the same objection is true of the dolphins that hunt cooperatively with humans—this might be a strategy that each dolphin learns on its own, reinforced by the fact that it receives a food reward after performing certain behaviors. Some have argued[173,174] that environmental variables, not social learning or culture, might best explain feeding specializations (e.g., kerplunking, beach hunting, fish whacking, bottom grubbing) in different populations of cetaceans.[175,176,177,178,179] And while the examples of killer whales teaching their young calves to beach themselves is a popular idea in the media, many scientists have pointed out that the evidence that this is truly teaching (and thus underpins the cultural aspects of this behavior) is mostly weak and/or anecdotal (i.e., still awaiting solid documented evidence that it fits the definition of "teaching").[180,181,182,183] A far better example of social learning via teaching in dolphins comes from observations of spotted dolphins in the Bahamas, where dolphin mothers appear to purposefully slow their swim speed when chasing fish so as to allow their calves the opportunity to learn proper hunting techniques, as well as toying with their prey in what appears to be an attempt to provide the calves with an opportunity to practice their prey-capture skills.[184] It is possible of course that slower chasing is the result of the mothers being distracted by their calves, although this is not likely the best explanation. It has also been suggested that dusky dolphins learn to find seasonal hunting grounds via the transmission of knowledge between unrelated animals, a form of oblique cultural transmission and social learning.[185]

A debate has been raging as to whether or not sponging, a seemingly clear-cut case of culture in dolphins, is a behavior that is exclusively (or primarily) a result of social learning.[186,187] Because sponging is correlated with specific habits (i.e., deep water channels), as well as occurs in individuals that are closely related, it is possible that both genetic, and environmental/ecological factors might be involved.[188] Although it might be too soon to declare a winner in this debate, research from the last few years seems to be showing that social learning is perhaps the better explanation as to why Shark Bay dolphins engage in sponging (and possibly other types of foraging) behavior.[189] Still, because foraging is observed in such a small population, there is some debate as to whether or not it's widespread enough in the Shark Bay population to warrant the moniker of "culture."[190] On the other hand, the fact that sponging seems confined to a small group of animals that, while solitary when engaging in sponging, spend more time socializing with other spongers (as opposed to non-spongers) is reminiscent of one of the definitions of human culture involving group identity and shared norms.[191] As it stands, because it is so difficult to both define culture and prove the existence of social learning in wild dolphin populations, a debate will likely rage on as to how appropriate it is to suggest that dolphins have culture. Many of the examples cited so far strongly suggest that culture/social learning is responsible, but, as Stan Kuczaj pointed out, "the suggestion of social learning is not the same as the demonstration of social learning."[192]

As far as uniqueness goes, culture as defined as "information or behaviour acquired from conspecifics through some form of social learning" is certainly not exclusive to dolphins. For example, ants[193] engage in a form of social learning whereby a "teacher" ant (i.e., the ant aware of the location of a food source) modifies its behavior (e.g., slows its running, stays in close contact, waits) in order to teach a follower ant the location of a food source.[194] These experimental results are perhaps stronger evidence of teaching in ants than in dolphins. In a controlled experiment with meerkats, naïve animals were able to learn novel techniques for acquiring food by interacting with animals that had learned the new technique.[195] These findings piggyback nicely on the discovery that meerkats in the wild actively teach their offspring hunting techniques by giving them the opportunity to interact with live prey.[196] The generally non-social red-footed tortoise is able to learn how to avoid obstacles and find a food reward by observing their peers doing so.[197] Social learning in fish seems rather widespread,[198] with fish learning, among other things, anti-predator behavior, predator recognition,[199] mating site location,[200] and migration routes by interacting with their peers. Lemon sharks can solve an experimental task more quickly by watching and interacting with conspecifics who have already learned to solve the task.[201] And, of course, chimpanzees display a wide range of socially learned behaviors, including tool use and grooming behaviors, which can vary from population to population.[202,203,204] Capuchin monkeys are perhaps even better candidates than great apes for exhibiting social traditions/culture.[205] But it's vocal dialects in birds that are perhaps the strongest examples of culturally transmitted behavior in any non-human animal species to date.[206,207] Of course, these findings from other species are subject to similar kinds of objections as the dolphin literature as far as definitions, methodologies, and interpretation is concerned. So what do these animal culture wars reveal about dolphin intelligence? It's difficult to say. We can draw the line between animal versus human culture wherever we would like, but we are still left with the problem of whether or not it's appropriate to consider having "human-like" culture a sign of intelligence (and thus a need for "special considerations")[208] or not.

5 Dolphinese

Animals have surprisingly rich mental lives, and surprisingly limited abilities to express them as signals.[1] *W. Tecumseh Fitch*

The prevailing theory about language at this moment in time is that no species other than *Homo sapiens* is in possession of a communication system that fits a linguist's definition of language. This is not a particularly controversial idea in the cognitive sciences, as one could gather by reading though the popular writings of leading scientists.

STEVEN PINKER: "Language is obviously as different from other animals' communication systems as the elephant's trunk is different from other animals' nostrils."[2]

STEPHEN R. ANDERSON: "When examined scientifically, human language is quite different in fundamental ways from the communication systems of other animals."[3]

TERRANCE DEACON: "Languages evolved in only one species, in only one way, without precedent, except in the most general sense."[4]

DEREK BICKERTON: "Everything you do that makes you human, each one of the countless things you can do that other species can't, depends crucially on language. Language is what makes us human."[5]

MICHAEL CORBALLIS: "This is not to say that nothing in the communication or actions of other animals bears on human language, but it is clear that the gap between human and animal communication is very wide indeed."[6]

NOAM CHOMSKY: "Human language appears to be a unique phenomenon, without significant analogue in the animal world."[7]

This isn't an oversimplified summary of a complex topic that these experts have repackaged for the general public. It is also expressed, equally as succinctly and straightforwardly, in the peer-reviewed literature, like this opening sentence from a 2011 article in the journal *Animal Cognition* describing the structure and function of chimpanzee gestures: "It is a truth universally acknowledged that the greatest cognitive difference between humans and other animals lies in the use of language."[8]

The phrase "a truth universally acknowledged,"[9] while nothing out of the ordinary for a Jane Austen novel, is a fairly powerful sentiment to be expressed in a scientific journal. The fact that the authors, reviewers, and editors of this article consented to publish this phrase speaks to how well established is the fact that language is fundamentally different to animal communication systems. You will be hard pressed to find a working scientist willing to argue that the totality of a given species' communication system (i.e., not just certain properties of it) is analogous, let alone homologous to human language.[10]

So is that it? Case closed? Dolphins don't have a language? No, this is the point in the story where the intelligent dolphin myth makes its grand entrance with tantalizing promises of dolphinese and the inevitability of interspecies communication. For many, dolphinese is the grand exception to the "truth universally acknowledged" that animals do not possess language.

Since John Lilly's initial proclamation that dolphins have a language, and dogged assertion that humans will one day hold a conversation with dolphins once we've learned to decipher it,[11] dolphinese has enjoyed a devout and vocal fan base. The tension between skeptical scientists and dolphinese proponents has been festering for nearly forty years. As early as 1972, mainstream science had already tentatively come to the conclusion that dolphin language was unlikely to exist, but when scientists went public with their views on this subject at the time, the backlash and "unbridled anger" they experienced from the public left them dumbfounded.[12] The rise of the environmental movement in the US, the passing of the Marine Mammal Protection Act, and the publication of Lilly's books *Man and Dolphin*,[13] *The Mind of the Dolphin*,[14] and *Communication Between Man and Dolphin*[15] meant that the 1970s was the wrong decade for skeptical scientists to begin a dialogue on why the public should tone down its rhetoric on dolphin language.

Today, the concept of dolphinese continues to crop up in the popular press. The nebulous idea of dolphin language is often alluded to by headline copy editors in a bid to hook the reader, with the meat of the story usually lacking a discussion of what language is, or whether or not it's appropriate to label dolphin communication as such. For example, *Discovery* magazine's article titled "Learning the Alien Language of Dolphins"[16] and *Wired Science*'s article "Dolphins May 'Talk' Like Humans"[17] are simply referring to dolphin language in the headline as a trope, without actually suggesting that dolphin communication is equivalent to human language anywhere in the main body of the text. These headlines are more poetic than informative, with the terms *language* and *talk* being used as a metaphor for *communication* and *communicate*. Bizarrely, FoxNews ran the headline "Australian Researcher Partly Decodes Dolphin Language" for an article discussing researcher Liz Hawkins' study of bottlenose dolphin whistles when she is quoted in the very same article as saying "a specialist in linguistics would not call this a language."[18] But even when the term *language* is avoided, the media is still keen on hyperbole, with Discovery News claiming that there is a "growing body of evidence that dolphins possess one of the most sophisticated communication systems in the animal kingdom, perhaps even surpassing that of humans."[19]

I should not be too harsh on the media for failing to emphasize the fact that mainstream science considers dolphinese to be a fanciful idea, especially considering how passionately dolphinese supporters continue to proselytize Lilly's original beliefs. "It is known that the dolphins (and presumably the other Cetacea) have a complex language—with up to a trillion 'words' possible," asserts the Hawaiian-based Sirius Institute, an organization founded by a former colleague of Lilly.[20] Global Heart, Inc., a dolphin research organization with the goal of expanding "communication between dolphins and humans" disseminated a press release in November 2011 announcing "The Discovery of Dolphin Language."[21] With stories like these crossing journalists' desks, the asymmetry that exists within the scientific community as to the existence of dolphinese (i.e., the overwhelming majority would not classify dolphin communication as a language) can be overshadowed by a journalistic attempt at balance. This leads to a situation where "The Discovery of Dolphin Language"—a finding that is not based on any peer-reviewed research—stands toe to toe in the media with the mountain of evidence suggesting that it just ain't so.

To be perfectly fair, mainstream scientists, myself included, are also guilty of muddying the waters when it comes to the question of dolphin language. Occasionally, when taking shortcuts to describe the science, we use turns of phrase that could suggest the existence of dolphinese, or otherwise make the distinction between language and animal communication fuzzier than it is. When Carl Sagan wrote of "dolphin language"[22] and "whale language,"[23] or Philip Lieberman stated that "animals have animal language,"[24] they were not implying that animal communication systems are equivalent to the linguistic/philosophical definition of a *natural language* (i.e., human language).[25] They are simply using an artistic figure of speech to describe animal communication science to the public. Some scholars even substitute the phrase "animal language" for "animal communication" for reasons of style, even when explicitly arguing that human language is unique.[26] The unintended consequence is that it reinforces the belief that dolphin communication truly is equivalent to human language in the Lilly sense.

With a powerful and well-established myth to contend with, and mixed messages coming from a variety of sources, it's understandable why the public and mainstream science are of two minds when it comes to the topic of dolphin language. To appreciate why scientists are so confident in stating that dolphin language does not likely exist requires that we understand what a linguist's definition of language is. We can then compare what is known of the natural communication system of dolphins to this definition. If I do my job as a spokesperson for mainstream science properly, by the end of this chapter you will agree with the experts that dolphin communication is a fascinating subject to study, but there is no convincing evidence that it should be classified as language.

What is Language?

The term *natural language* refers to human languages, like English or Chinese. Scientists and philosophers use the term natural language as a means of differentiating between language proper and other communication systems that receive the colloquial moniker of language, like computer programing languages (e.g., C++, PHP, Java), the language of mathematics (i.e., mathematical notation), or the unmistakably poetic *language of love*. A definition of natural language that I often use is as follows:

> An arbitrary set of learned symbols (usually vocal) shared by a group, consisting of discrete elements that are combined following the rules of a grammar system to represent limitless concrete and abstract meaning.

Using a finite set of phonemes (i.e., the smallest meaningful unit of sound in a language) and a set of rules for combining them (i.e., syntax/grammar), natural languages generate novel phoneme combinations resulting in words, phrases, and sentences. Linguists refer to this as a *discrete combinatorial system*. In theory, there is no limit to the number of novel combinations that a discrete combinatorial system can generate. There is also no limit to the number of concepts or ideas that a language can map onto these symbol combinations, resulting in the ability to express any concept imaginable. If the human mind can think it, there should be a way to represent it using language. Because of these properties, natural language is much more than just a *communication system*. Traffic lights (a system of communication but not a language) also make use of arbitrary symbols (colors) that convey a message (stop, caution, go) using an agreed upon system.[27] But traffic lights, C++, and love are not *used to represent limitless concrete and abstract meaning*. Unlike language, the symbols in most communication systems are not recombined in novel ways to generate novel concepts—they are closed, inflexible systems with a finite set of meaningful outcomes. It is the property of *limitless expression* that is the crucial difference between natural language and *animal communication systems* (which I will subsequently label using Derek Bickerton's acronym of ACS).[28]

It's not the case that linguists have created this definition of language in a homocentric attempt to prevent ACS from attaining the label of *language*. If scientists were to find an ACS that demonstrated a capacity for limitless expression, they would gleefully consider this a natural language. But they simply have not. Limitless expression is the end product of natural language, and it is very likely that this outcome is unique in the animal kingdom. But how the human brain accomplishes this, to what extent the underlying cognitive abilities are shared by other species, and how and why these abilities evolved in the first place is not at all clear.

What is the explanation as to how our early primate ancestors transformed their ACS into natural language? The debate as to exactly what unique evolutionary circumstances and cognitive properties were necessary for language to have evolved is unlikely to be settled anytime soon, if ever. The language evolution debate itself has a long and acrimonious history. Tired of the endless squabbling and lack of progress on the topic, the Linguistic Society of Paris banned papers on the subject of language evolution in 1866. One hundred and fifty years later, it's obvious that this ban has long since been lifted: I am looking at a stack of books and papers on my desk from modern scientists offering their own take on how and/or why language evolved: the ability to think with symbols (Terrance Deacon);[29] recursive thought (Michael Corballis);[30] niche construction theory (Derek Bickerton);[31] gestures and cooperative communication (Michael Tomasello);[32] changes to speech production anatomy (Philip Lieberman);[33] vocal grooming (Robin Dunbar);[34] natural selection of ACS traits for language (Steven Pinker);[35] the vital role of kin selection (W. Tecumseh Fitch);[36] a single fortuitous mutation (Noam Chomsky).[37] The correct answer could involve all of these things, or none of them.

Fortunately, it's not necessary to have a firm grasp on how language evolved in order to understand why dolphin communication is not a language. We don't need to have identified any magic bullet(s). We've already identified the fundamental difference between the two systems that looms large over this discussion: language is capable of limitless expression and ACS is not. And we have many lists of cognitive attributes that we know for sure are involved in generating this outcome, even if we don't fully understand how they interact in the brain to accomplish this, or why they evolved. Scholars have been compiling lists of essential components of language for decades, with the most famous and influential being Hockett's thirteen design features of language.[38] But the modern study of animal cognition has made one thing clear: any cognitive attribute you add to a list like this can and has been found to exist in some form in a non-human animal species. Nonetheless, and this is the key issue, only humans appear to be creating language from this assembly of semi-universal cognitive components. As we look at dolphin communication, we will discover that dolphins possess many of the cognitive attributes one would hope to see in a species that possesses language, but that they too fall short of possessing something akin to a natural language.

Ten Essential Ingredients of Language

In order for dolphin communication to be considered a language, it would need to fit a linguist's definition of natural language. If you will indulge me, I would like to shoehorn my own list of ten attributes necessary for language into this discussion so we can compare humans and dolphins in a more formal way. This list is cobbled together from the ideas proposed by the big thinkers I've been referencing (e.g., Hockett, Chomsky, Fitch).[39] It includes concepts that scientists have identified as being absolutely vital to both the production and comprehension of a communication system that we could label a natural language.

1. Limitless Expression—the ability to express any idea or concept; the essential property/outcome of language.
2. Discrete Combinatorial System—a syntactic/grammatical system allowing the mind to combine a finite set of small meaningful units to generate infinite meaningful combinations.
3. Recursion—the ability to embed syntactic structures inside like syntactic structures *ad infinitum*; the workhorse of the discrete combinatorial system.
4. Special Memory for Words—a memory system that is especially attuned to storing the meaning of discrete symbols (usually vocal symbols) like words.
5. Displacement and Mental Time Travel—the ability to convey information to the listener about events that one cannot see, and/or occurred in the past or might occur in the future.
6. Environmental Input Required—the idea that no language can be generated spontaneously; all properties of language (e.g., grammar, phonetics) must be learned via interaction with skilled speakers.
7. Arbitrariness—the idea that the symbols used in a language bear no discernible resemblance to the things they describe.

8. Freedom from Emotion—linguistic concepts are not restricted to symbolic representations of internal (emotional or motivational) states.
9. Novelty Generation—the ability to both learn and create new ideas and represent them with new symbols (i.e., neologisms).
10. Social Cognitive Aptitude—the ability to intentionally convey information to a listener by ascertaining what they might or might not already know. This requires an ability to mind read (i.e., have a theory of mind) to some extent.

I have ignored traits of language described by Hockett, Hauser, Chomsky, Fitch, Lieberman, and others relating to specializations in sensory motor systems. Evolutionary changes to speech production or sound processing anatomy never struck me as particularly vital to a modern definition of language. A descended larynx is certainly important to language sound production, and its evolution might well have been closely intertwined with the cognitive muscle necessary to create language (e.g., vocal imitation), but for the current human phenotype, it's a non-issue. The fact that human sign languages make use of modalities and perceptual systems foreign to most language systems (e.g., vision instead of audition, gestures instead of vocalizations) demonstrates that the vocal tract and auditory system are merely used by, but not vital to, language production/comprehension in modern *Homo sapiens*.

Every single one of these ten ingredients can be found in the communication systems of non-human animals to some degree. There simply is no fundamental break in continuity between the cognitive abilities of human and non-human animals—we are all essentially basing our communication systems (and other behaviors) on the same basic cognitive machinery. This is in keeping with Darwin's sentiment that "the difference in mind between man and the higher animals, great as it is, certainly is one of degree and not of kind."[40] But differences in how humans use this machinery are what have created the gulf between ACS and language, something akin to a "functional discontinuity" in cognition according to Derek Penn, Keith Holyoak, and Daniel Povinelli in their 2006 article in *Behavioral and Brain Sciences*.[41] The suggestion that human and animal cognition is separated by a gap so wide that perhaps Darwin made a "mistake" in describing it as a difference in degree evoked a passionate response from others in the science community (i.e., a "rush to defend Darwin's honor"[42]). Dolphin researchers Louis Herman, Robert Uyeyama, and Adam Pack submitted a reply citing examples of dolphins' skill at symbol comprehension as a challenge to the authors' characterization of cognitive disparity.[43] Much of the lively debate that followed the publication of this article focused on how we should describe the gap between human and animal cognitive (and linguistic) abilities: is it a difference of degree, a difference in kind, a functional discontinuity, qualitative phenomenological differences, strong versus weak discontinuity, a sharp divide, etc.? But getting bogged down by a jargon debate is unproductive. There *is* a disparity between ACS and language, we just need a way of quantifying it without resorting to qualitative (and potentially divisive) descriptors which are, at the end of the day, simply a matter of taste.[44]

In order to facilitate a quantitative comparison, I propose a method of rating the extent (i.e., degree) to which each of the ten essential ingredients of language have been found to exist in dolphins or other species. It is clearly foolish to simply state that a trait has or has not been found, insofar as many of these aptitudes arguably exist to some extent in an ACS, but not to the degree found in a natural language. In lieu of a yes/no rating, a Language Rating Scale from 0 to 5 will be used for each ingredient, with 0 representing complete absence, and 5 equating full-blown aptitude. A total rating of close to 50 would indicate that dolphins are close to possessing a natural language. I base each rating on studies of the natural communication system of dolphins, experiments to teach dolphins an artificial symbolic communication system, and general research into dolphin cognition.

The Dolphin Language Rating

Limitless expression | rating = 0

After Lilly's proclamation that dolphins possess a language, it was natural that early researchers presupposing the existence of dolphinese would begin their search for evidence by looking for the defining characteristic of language within dolphins' vocal communication system: the ability to communicate novel, abstract information. The first researcher to test for this ability was Jarvis Bastian, a psycholinguist from the University of California, Davis. In the 1960s, Bastian designed an experiment in which dolphins were required to inform each other about an (arbitrary) event using whatever communicative skills they had at their disposal in order to obtain a food reward.[45] Two dolphins—a male named Buzz and a female named Doris—were placed in the same pool that was divided in half by a mesh net. They were able to both see each other and hear each other's vocalizations. The dolphins were trained to press a paddle that corresponded to the state of a light positioned above their pool—either flashing or steady. They were only rewarded if both dolphins pressed the correct paddle, and only if Buzz pressed his first. The two dolphins had little trouble learning this task (i.e., small potatoes as far as dolphin cognition goes). In the next phase of the experiment, an opaque barrier was placed between the two pools, so that the dolphins could no longer see each other, but could still hear each other's vocalizations. Importantly, Buzz was now no longer able to see the light—only Doris knew whether or not it was flashing or steady. So it was up to Doris to "tell" Buzz about the light, so that he could press the correct paddle and they would both obtain their food reward. Much to the researchers' delight, this is exactly what appeared to happen. While positioned in front of the paddle corresponding to the state of the light, Doris emitted a series of vocalizations, and Buzz pressed the correct paddle in his pool on nearly every trial. The tentative conclusion was that Doris had been communicating arbitrary novel information to Buzz (i.e., the state of a light) using her vocalizations.

Bastian was not satisfied with this experiment, and, as is the hallmark of science, began implementing strict controls in subsequent experiments to ferret out potential confounding factors that might alter his provocative conclusion. He found that Doris produced the same vocalizations regardless of whether or not Buzz could see the light, and even if Buzz was not in the pool.[46] This challenged the idea that Doris was intentionally transmitting information to Buzz about the state of the light. The more parsimonious explanation for the original finding was that Buzz was using other information to determine which paddle he needed to press. The final conclusion was that Buzz was able to localize where Doris was in the pool (i.e., which paddle she was positioned in front of) by listening to her vocalizations. He then chose the correct paddle by simply being able to differentiate between Doris being farther or closer away when she vocalized, not via linguistic information found in her vocalizations.[47,48] Researchers at Dolphinarium Harderwijk in the Netherlands replicated Bastian's study and obtained almost identical results to that of Bastian.[49] After analysis of the dolphins' vocalizations, the Harderwijk researchers determined that there was no difference in the structure of the vocalizations emitted by the dolphins for each for the two conditions (i.e., light flashing or steady), and that it was the position of the vocalizing dolphin that allowed the listening dolphin to make the correct choice.[50] Unfortunately, the Discovery Channel/ Animal Planet[51] and a handful of authors[52] chose to popularize the conclusion from Bastian's initial experiment (i.e., dolphins can communicate abstract ideas), ignoring the fact that a host of other scientists, including Bastian himself, later insisted that this conclusion was unequivocally wrong.

The only possible experimental evidence that limitless expression might still be an existing phenomenon for dolphin communication stems from two publications hailing from Soviet scientists. These publications are often cited by dolphinese proponents as scientific proof of the "open vocabulary" of dolphin communication,[53] although they've been largely ignored by mainstream science. A brief explanation as to how these articles entered the public (although not scientific) discourse is necessary.

In 1989, two years before the collapse of the Soviet Union, a conference was held in Rome, Italy, concerning the topic of the sensory abilities of cetaceans. A number of scientists from the Soviet Union attended, presenting their findings in English—a rare occurrence in the years before Mikhail Gorbachev introduced the policy of glasnost. Western scientists were, for decades, unaware of the extensive research being conducted on dolphin cognition behind the iron curtain. As is typical of this kind of workshop/symposium, the authors were asked to transform their presentations into articles to be included in the conference proceedings appearing in book format. It is important to point out that although these articles were edited by experts in the field, conference proceedings as a rule are not subject to the kind of peer review that one finds in academic journals, thus the authors' methods, statistical analyses, and discussion points were never critically evaluated by a jury of their peers.

The first article, with lead author Alexander V. Zanin, describes an experiment with dolphins from the Black Sea that was a modified version of Bastian's initial experiments.[54] In their introduction, the authors state that it is their opinion that dolphins' natural communication systems can report complex arbitrary information. They suggest that there is "indirect corroboration of this standpoint such as some features of the dolphin behavior in the wild and captivity," although they do not mention or cite what behavior(s) they are referring to. Because of the failure of previous experiments (specifically Bastian's work with Doris and Buzz) to provide experimental evidence of this ability, the authors' aim was to design an experiment that could conclusively show that dolphins can report arbitrary semantic information via their vocalizations.

The researchers set up an experiment in which one dolphin was given a visual signal to choose one of two balls (i.e., large or small) suspended in its tank. If this dolphin and a second dolphin in an adjacent tank (that was able to hear but not see the other dolphin or the visual signal corresponding to the correct ball) both chose the correct ball, they were both given a food reward. The dolphins would have been able to hear each other's vocalizations. Correct choices were attributed to dolphin A sending arbitrary semantic information (i.e., which of the two balls dolphin B needed to choose) via its vocalizations. Unfortunately, there is no description of controls to eliminate the possibility that simple acoustic cues like those in the Bastian experiments (e.g., different spectral characteristics of vocalizations produced when positioned in front of the large ball as opposed to the small ball) contributed to dolphin B learning the correct choice. Like the Bastian experiment, this experiment cannot be considered an appropriate test of dolphins' ability to transmit arbitrary semantic information via their natural signals.

The second article appearing in the proceedings by Vladimir I. Markov and Vera Ostrovskaya[55] describes the use of a "rank distribution method" (i.e., Zipf's law) to show that dolphin vocalizations contain semantic content, and a calculation showing that the distinct sound units dolphins produce could be combined to produce 10^{12} different kinds of hierarchically organized signals. If indeed dolphins could map semantic information onto these different signals, they just might satisfy the criterion for limitless expression. However, to make this claim, the authors must make a direct appeal to the Building Blocks Fallacy (described later). Moreover, in the light of our current understanding of the use of information theory (e.g., Zipf's law) to study animal communication, we now know that the methods outlined in this article for attributing semantic content to dolphins' signals are deeply flawed, a subject that will be covered in more detail in the section on information theory later in this chapter.

These authors did not publish on these topics again after the collapse of the Soviet Union in 1991. A Russian colleague of theirs, Mikhail Ivanov from St. Petersburg State University, took inspiration from this research and published on his experimental design for eliciting potentially complex communicative behavior in captive dolphins.[56] However, unlike the Zanin experiments, these were not direct tests for the transmission of arbitrary, semantic information. Thus, these two articles from *Sensory Abilities of Cetaceans* stand alone as the only scientific publications ever published (that I know of in English) having suggested that experimental evidence exists for the capacity for limitless expression in dolphins' natural communication systems. Given the problems I've just outlined with these studies, it's perhaps understandable why the mainstream scientific community was not persuaded to alter their conclusion that dolphin communication lacks this all-important ingredient of language.

These two Soviet experiments together with the Bastian and Harderwijk experiments were the only attempts to test for dolphins' ability for limitless expression. Were these experiments enough to draw the final conclusion that this ability was absent from dolphin communication? For generations of scientists, the answer was yes. I was once quoted in an interview as saying that "essentially [dolphins] do behave in complex and interesting ways, but there are no great mysteries in what they do that can only be answered with language."[57] What I meant by this is that when you look at how dolphins behave in the wild and in captivity, there is nothing about their behavior—even in their demonstrably complex behavior—that cannot be accomplished by an ACS that lacks the ability to transmit limitless (or even complex) semantic information. Dolphins' ability for coordinated group hunting and maintaining intricate fission–fusion social structure can be mitigated by normal modes of animal communication. Indeed, moray eels[58] and elephants[59] accomplish these feats without anyone suggesting that their communication systems are capable of limitless expression. Because there is no need to posit language as an explanation for dolphin behavior, and because dolphins failed these initial experimental tests for limitless expression, this line of research has fallen out of favor with modern researchers.

To be clear, there have been, and still are, great mysteries of animal communication that drive similar research. As early as the 18th century, scientists suspected that bees were able to communicate abstract information about the location of food sources with their waggle dance. How they accomplished this remained a persistent mystery until Karl von Frisch identified the mechanisms of the waggle dance. And today we face the mystery of how prairie dogs are able to communicate information about the size, shape, and color of objects in their environment via their vocalizations—perhaps the most startling example of semantic information transmission in any animal communication system ever discovered (and worthy of a rating of at least 2 for Limitless Expression on the Language Rating Scale).[60] But alas for dolphins, no mysterious behavior akin to bees' uncanny habit of finding flower patches or prairie dogs' ability to inform each other of the color of my shirt has ever been observed. There is nothing in dolphins' behavior that is so different to the behavior of other species that leads scientists to believe that dolphins are unique in having the potential for limitless expression hidden within their vocalizations.

Discrete combinatorial system | rating = 1

Some of the first experiments to search for the basic properties of dolphinese were conducted in the early 1960s by John J. Dreher. Dreher was a mathematical linguist and acoustics engineer with the Lockheed Aircraft Corporation contracted by the US Navy to study dolphin vocalizations. Dreher and colleagues recorded sounds produced by captive animals at Marineland of the Pacific, as well as wild dolphin groups, in order to identify vocalizations that could be correlated with behavior—an attempt to ferret out the presence of "words," a fundamental component of a discrete combinatorial system.[61] The basic idea driving this line of research was that dolphin whistles, which vary in the rate and degree to which the tone rises and falls (i.e., whistle contour), were analogous to individual words or phonemes. If researchers could begin to associate specific whistle-words with specific behaviors, they could begin to "decode dolphinese." Dreher's results provided both an early catalog of different dolphin whistles, and attempts at mathematically calculating their potential information content.[62,63] Other early researchers, including John Lilly, created similar catalogs of whistles and pulsed sounds produced by dolphins.[64,65] Some scientists at this time were already critical of the widespread assumption that dolphin vocalizations were linguistic in nature, with one scientist replying to Dreher's article on dolphin communication in the book *Whales, Dolphins and Porpoises* (compiled from contributions of the First International Symposium on Cetacean Research held in 1963) with the following prophetic assessment:

> I shudder every time I hear the word "language" used in the context in which it is being used here today. Dr. Dreher, today, and others on previous occasions, have found it necessary to explain that when they employ the term "language" in reference to porpoise [dolphin] sounds, they are using it in a difference sense from the customary meaning of this word as used from time immemorial. If this practice continues, we will have one term for two very different meanings. For a group of experts on communication, this certainly is not very efficient. It seems to me that the broader word "communication" conveys the meaning of everything that I have heard here today on the subject of cetacean sounds.[66]

Later researchers confirmed that these early studies of whistle variation provided no evidence of the kind of diversity needed to expect that dolphins were combining their sounds in a manner consistent with a discrete combinatorial system.[67] The cataloging of dolphin whistles has continued to this day (although with different goals in mind), with sophisticated computerized methods of lumping and splitting different whistle types based on numerous whistle properties.[68] We've even reached a point where our ability to categorize whistles has been shown to match how dolphins themselves categorize whistles.[69]

Correlating whistles or whistle properties to observable behaviors is a useful approach to studying dolphin communication. Whistles are likely used to transmit information about what sort of behavior an individual dolphin is engaging in—from foraging to socializing.[70] Signature whistles transmit information about the identity of the whistler. Whistles are used for maintaining contact, cueing the direction of travel, and alerting conspecifics to the presence of food.[71] Changes in whistle duration and the number of inflection points (i.e., where the whistle contour reverses direction) might be used to convey different levels of emotion (e.g., alertness, fear, aggression).[72] Burst pulses and other pulsed sounds are also used as cohesion calls, for individual recognition, during aggressive encounters, and as food calls. All of these functions fall within the domain of ACS, and are typical of non-linguistic animal communication. The idea that whistles or pulsed sounds are being combined by some kind of syntactic system to form semantically rich sentences has been absent from the scientific literature for decades (with the exception of the Soviet studies described earlier). In a chapter on dolphin communication from the 2009 book *Vocal Communication in Birds and Mammals,* dolphin communication expert Vincent Janik from the University of St. Andrews dedicates just one single sentence to this topic: "There is little evidence for syntactical rules in delphinid vocalizations, even though they have been found to comprehend relatively complex syntax in artificial sign systems."[73] Janik has, however, suggested that there is "a large potential" for syntactic structure to be found in dolphin clicks and pulsed calls,[74] although syntactic structure need not be mapped onto a semantically rich communication system (as is the case for birdsong) leading to a discrete combinatorial system. There is, however, some evidence that dolphin whistle repertoires might be sufficiently diverse and produced in a non-random manner to suggest that they have the capacity to transmit large amounts of information—an idea first explored by Dreher. But, as the discussion of information theory later in this chapter will make clear, this is not evidence of a discrete combinatorial system.

The reason dolphins achieved a rating of 1 for discrete combinatorial system is not because of anything found in their natural communication system, but because of their ability to use "artificial sign systems," as Janik pointed out. In the Kewalo Basin studies, Akeakamai and Phoenix were trained to use artificial language systems comprising visual and acoustic signals (respectively). These signals were signs referring to objects, actions, and locations. Two or more signs could be combined in a sentence to convey unique instructions. Importantly, changes to the order in which these signs were presented would change the meaning of the instructions. A sentence consisting of the signs for *person surfboard fetch* presented in that order would mean "bring the surfboard to the person." But the phrase *surfboard person fetch* would mean "bring the person to the surfboard." Akeakamai and Phoenix were able to learn the difference as it pertained to word order and apply it to novel sign combinations, demonstrating that they had the ability to comprehend basic syntactic rules about combining signs. Although there is no evidence that dolphins apply this ability to their own vocalizations, the capacity for comprehending combinations of discrete elements to produce meaningful sentences exists in the dolphin mind. However, comprehension was limited to basic word order in relatively short (maximum five word) sentences. This constitutes only a tiny subset of syntactic properties we find in the discrete combinatorial system underpinning full-blown natural language (and perhaps warrants a heavily qualified definition of the term "syntax"),[75] which is why the Language Rating did not creep above 1. It should be noted that this skill is not unique to language-trained dolphins: Alex the parrot and Kanzi the bonobo (among other animal language prodigies) have shown their ability to both comprehend and produce utterances consisting of basic sign combinations.[76] And dogs' well-known ability to comprehend verbs paired with nouns when given as a command (e.g., *ball fetch*) in various contexts shows that dogs too understand a basic form of syntax involving sign combinations.[77]

Recursion | rating = 1

Recursion in the linguistic sense is the ability for grammar to embed structures within structures, using its own output as input. In the following example, recursion underlies the syntactic rule of *center embedding*, where a phrase can be embedded within another phrase *ad infinitum*:

> The rat ate the cheese
> The rat **that the cat chased** ate the cheese
> The rat that the cat **that the dog bit** chased ate the cheese[78]

Some have argued that the ability to generate recursive linguistic structures is both unique to the human mind and the essential property of language.[79] The only reference to dolphins being able to either produce or comprehend recursive structures was found in a study of word string comprehension by Akeakamai and Phoenix. The authors described the dolphins' ability to comprehend "recursive forms including conjoined constituents and conjoined sentences."[80] The dolphin Phoenix was given instructions consisting of two linked sentences, for example: *pipe tail-touch pipe over.*[81] This would indicate that she needed to first touch the pipe with her tail, and then jump over the pipe. Phoenix had little problem parsing the meaning of this syntactic structure, carrying out the correct action for eleven of fifteen novel conjoined sentences. However, the kind of recursion being described here falls under the category of tail recursion,[82] which involves tacking on constituents to the end of a phrase. Like center-embedding, this could result in sentences of infinite length (limited in practice by primary memory). But unlike center-embedding, the first phrase and second phrase—although given in the same sentence—are standalone phrases that do not reference each other; one does not need to remember the meaning of the first phrase to understand the meaning of the second. This is more like an iterative syntactic process, and is frequently observed in bird and primate communication where vocalization sequences are simply repeated.[83] There are arguments that iteration (and thus the kind of tail recursion comprehension displayed by Phoenix) might be the computational process responsible for the infinitely productive syntactic systems seen in human language,[84] which is why

I am happy to give dolphins a rating of 1 for Recursion. However, for many researchers, tail recursion is not considered the kind of recursive structure fundamental to a discrete combinatorial system (i.e., language).[85] Unlike humpback whales[86] and Bengalese finches,[87] dolphins have not yet been shown to comprehend or produce center-embedding or similar hierarchical syntactic structures, and none of these species appear to harness the power of recursion to create a discrete combinatorial system or limitless expression.[88,89]

Special memory for words | rating = 1

The human mind appears to have a rather impressive ability to store the distinct meaning of individual words. It's actually rather difficult to produce an estimate of the average vocabulary of language users for a variety of reasons: 1) how do we quantify the distinction between words one is able to produce and words one comprehends or recognizes; 2) where do we draw the line between a word and inflected versions and derivations of that same word (i.e., word families) which might have a different meaning (e.g., choose, chooses, chosen, chooser, choosy); and 3) what do we do with words one knows in different languages that are representing the same meaning (e.g., airplane, vliegtuig, avión)? Depending on how a researcher deals with these issues, vocabulary estimates for the average college-educated speaker (in their native language) range from 17,000[90] to 216,000.[91] Tacking on foreign or second languages one knows might well double or triple those estimates.

Although it's not correct that the concept of a word has a homolog in ACS, many have suggested that alarm calls could be considered analogs to words.[92] Alarm calls are vocal signals produced by a number of species (e.g., chimpanzees, vervet monkeys, meerkats, chickens) upon seeing a threat in their environment (e.g., a snake, a hawk). Upon hearing this signal, other animals (of both the same and different species) will take evasive action based on the type of urgency or predator associated with the specific call. This is a somewhat simplistic overview of the situation, as it is still hotly debated about how closely the signal is linked to its referent, and whether or not the animals form a mental representation of the specific threat to which an alarm call supposedly refers.[93] But even when adopting a liberal attitude toward analogy and assuming that individual alarm calls are functionally equivalent to words, the "vocabularies" of alarm call producing species are pitifully small when compared to humans—usually not exceeding six different call types.[94] Of course, the same alarm call can be modulated based on, for example, urgency or the distance of the predator in order to convey different information, so this is not a very fair comparison. Regardless, alarm calls are clearly not vocabulary rich.

Animals involved in artificial training programs are capable of learning a variety of different symbols with distinct meanings. The largest reported symbol repertoires include Kanzi the bonobo (over 200 lexigrams),[95] Rico the Border Collie (over 200 labels),[96] and Alex the parrot (over 150 labels).[97] As for dolphins, Akeakamai and Phoenix were able to learn between forty and fifty distinct symbols and their meanings.[98] This is certainly impressive, and deserves at least a 1 for Special Memory for Words. But if we put this in perspective, we see that Chaser the Border Collie is the true star of the animal vocabulary Olympics, not Ake or Phoenix. Chaser learned the labels of 1,022 different objects—the most labels ever learned by a nonhuman animal.[99] But even Chaser has a vocabulary that is an order of magnitude (or two) less than the average human. Non-human animals simply do not have a memory for symbols to the extent that humans do. Also, keep in mind that a human can combine words into phrases and sentences that differ in meaning to any individual word in their vocabulary (e.g., the concepts of "dog," "dog-food," and "dog-food-dispenser" all have different meanings). ACS cannot do this (as far as we know), and even language-trained animals only do this to a limited extent (i.e., by combining at most two or three words to form novel concepts).

Displacement and mental time travel | rating = 3

There are two conceptual issues integral to the flexibility and openness of a natural language: 1) the ability to refer to objects and events that are not immediately perceptible (e.g., happening out of sight), and 2) the ability to refer to objects or events that existed in the past or might exist in the future. These concepts are respectively known as displacement and mental time travel. Research at Kewalo Basin has provided us with evidence that these abilities exist to some extent in dolphins. Akeakamai was able to report (by pressing a yes or no paddle) whether or not a specific object was present in her tank. When asked, for example, "is there a Frisbee?" when there was no Frisbee in sight, Ake almost never failed to provide the correct answer (i.e., "no").[100] This would require Ake to understand that an object that she knows exists (based on her mental representation of the object and past experience interacting with it), but that she could not see at present, nonetheless continued to exist in a location other than her pool. And importantly, she was able to communicate this information—satisfying the criteria for displacement. There is also the possibility that dolphins use their natural communication system to refer to objects that they cannot immediately perceive. It has been shown that dolphins occasionally imitate the unique signature whistles belonging to their social partners, effectively referring to an object (i.e., another dolphin) they might not be able to see.[101] It has been shown that it is the contours of signature whistles that convey identity information, not just the unique qualities of the individual caller's voice.[102] But it remains to be seen whether the function of signature whistles truly is referential in nature, or if producing a social partner's whistle is simply another version of a shared contact call, without the signaler having a mental representation of the specific individual that is associated with a particular whistle.[103] Regardless, dolphins certainly have this mental capacity.

Two Kewalo Basin dolphins, Elele and Hiapo, were shown to remember and thus mentally represent not just objects, but events that have happened in the past. The two dolphins were able to follow commands such as "repeat the previous behavior," or "perform any behavior except one that has been recently performed." By successfully performing these behaviors, Elele and Hiapo clearly had the ability to represent actions that had occurred in the past—an attribute of mental time travel that was discussed earlier. The results of these series of experiments which show that dolphins can represent (and thus communicate about to some extent) past events, and refer to objects and events they cannot immediately perceive, result in a Language Rating of 3. It fails to reach 5 because of the fact that we've not seen the ability to represent future events in this manner, and it has yet to be shown if these abilities manifest themselves in dolphins' own communication systems.

Environmental input required | rating = 3

Children acquire language via interaction with proficient language users (adults and older children), tacitly learning the rules of phonology and syntax (and other aspects of language) over the course of many years. To produce a fully skilled native speaker of a natural language does not require explicit language instruction, but it does require long-term language exposure. Whether or not this process is governed by a specific language acquisition device in the mind or by more generalized learning modules is still a matter of debate. Nonetheless, a protracted period of environmental input is fundamental to the acquisition of a natural language.

Vocal learning, "the ability to acquire vocalizations through imitation rather than instinct," is rare in the animal kingdom, seen only in humans, songbirds, hummingbirds, bats, parrots, seals, and cetaceans.[104,105] Dolphins have been known to be particularly skilled vocal mimics,[106] easily reproducing novel artificial whistles.[107] This mimicry skill is fundamental to the development of the communication system of young dolphins. As calves begin to develop their whistle repertoire—particularly signature whistles for bottlenose dolphins—environmental input is essential. Young dolphins will learn to imitate the whistles of their tank mates (including but not always their mother's whistles), as well as artificial whistles generated by researchers, and even trainers' whistles.[108,109] The first year of a bottlenose dolphin's life appears to be a critical period for the formation of a signature whistle.[110] However, dolphins retain the ability to learn new whistles throughout their lives, sometimes altering their own signature whistles,[111] or otherwise updating their repertoire to reflect changes in social partners.[112]

Killer whale pods produce distinct vocal repertoires called dialects. These dialects are unique to a matrilineal group and are learned by calves during the first few years of life.[113] Unlike bottlenose dolphins, these calls are thought to primarily be passed down vertically—from mother to offspring. Killer whales have, however, been observed mimicking the unique calls of other killer whale pods,[114] and it is now thought that the development of pod specific dialects may involve—to some extent—the borrowing of features from the calls of "rival" pods.[115] Taken together, all of these findings suggest that dolphins are one of a rare subset of mammals that requires environmental input in order to acquire its natural communication system, yielding a Language Rating of 3. Unlike humans, however, there is no evidence that a lack of exposure to conspecific vocalizations for dolphins will result in an utter lack of skill in communicating with conspecifics—as it would for natural language (although this has not been explicitly tested). Thus, the Language Rating did not attain a full rating of 5.

Arbitrariness | rating = 4

The simplest interpretation of arbitrariness (first popularized by Hockett) is equivalent to the idea of non-iconicity. Unlike a symbol, an icon bears a resemblance to the thing to which it refers. An example of a non-arbitrary or iconic signal might be pantomiming drinking from a cup to stand in for the concept of drinking from a cup, or mimicking the sound of a honking horn to refer to a honking horn. So in the example of chimpanzee or vervet monkey alarm calls, the vocalization given when a snake or leopard is spotted is non-iconic insofar as it bears no resemblance at all to any physical or other characteristic of an actual leopard. Nor does it represent another possible interpretation of the call, like the suggestion to *run up a tree* or *take cover*. It's just a series of random sound vibrations that the monkeys have evolved to associate with the presence of a leopard and/or the behaviors one performs to avoid being eaten by one.

Arguably, every vocalization produced by a dolphin and used for communication is arbitrary in nature. Of course, there is no evidence that dolphin vocalizations—other than signature whistles—refer to objects or events, which makes it a bit difficult to discuss iconicity or non-iconicity in the way we can for (potentially referential) alarm calls. However, there is also no doubt that dolphins are able to associate arbitrary sounds and visual signs with objects, actions, and events, as evidenced by the Kewalo Basin body of research, which gives them a 4 on the Language Rating Scale for Arbitrariness.

Freedom from emotion | rating = 2

For centuries, the prevailing view has been that ACS is primarily (if not exclusively) used as a means of conveying information about the caller's internal physiological or emotional state.[116] Donald Griffin termed this the "groans of pain" or GOP interpretation of animal signaling.[117] To some extent, this is a fair portrayal of how ACS works for many species. A lion's roar, a pig's squeal, a cat's purr—these are all signals that convey important information about the animals' state of mind to a potential receiver. Humans display similar kinds of signals; blushing when embarrassed, screaming when frightened, or smiling when happy. Even within human language, various (usually involuntary) paralinguistic elements convey information about our emotional state, for example, rapid speech production at a higher pitch might indicate fear, whereas variable pitch production could indicate happiness.[118] But for ACS, there are plenty of examples of signals that, like human language, are communicating information that is not just about an animal's emotional state. Referential signals like meerkat alarm calls or baboon "move" grunts may contain affective information, but are based on a call that transmits information about a behavioral, social, or environmental context external to the caller's mind and independent from their emotional state.

Animals taught to communicate using signs that refer to objects and events are obviously capable of communicating about subjects other than their own emotional state. This includes Ake, Phoenix, and other language-trained dolphins. It is, however, still an open question as to the extent to which dolphins' natural communication systems are conveying non-affective (e.g., referential or semantic) information. Signature whistles might well be an example of a signal that, like a predator alarm call, conveys referential and affective information simultaneously. But what do we make of all the other whistles and pulses produced by dolphins? Most research in this area suggests that changes to vocalization patterns (e.g., whistle rate, amplitude, inflection points, duration),[119] including signature whistles, correlate with behaviors associated with arousal states or behavioral contexts;[120] the typical association between ACS and emotion. Even complex social situations involving coordinated hunting and bond maintenance appear to rely on affective as opposed to referential signaling. Even though we know dolphins are capable of representing non-affective information via arbitrary signals, the evidence suggests that their natural communication system does not avail of this ability, which results in a Language Rating of just 2.

Novelty generation | rating = 2

At its core, novelty generation involves the creation of new signs to stand in for objects, events, or concepts for which a group of language users does not yet have a sign (e.g., e-reader, cloud computing, photobomb). Given that there is no evidence that dolphin communication has distinct linguistic units that—like words—stand in for objects, events, or concepts, novelty generation is unlikely to be a part of their natural communication system. Dolphins are capable of learning new signs within an experimental paradigm (as in the case of dolphins spontaneously learning and imitating whistles that refer to objects),[121] and they can spontaneously comprehend the meaning of novel sentences using already established signs. During Diana Reiss' research in the mid-1980s involving creating an artificial whistle communication system for two dolphins, she noted one observation of dolphins spontaneously combining whistles they had learned for two objects (i.e., ball and ring) into a single whistle that they emitted when playing with both objects at the same time, which certainly resembles the idea of novelty generation.[122] Even though novelty generation is unlikely to exist for their natural communication system, this single experimental observation is enough to garner them a Language Rating of 2 for this ability.

Social cognitive aptitude | rating = 3

Many ACS signals are produced involuntarily or reflexively (as the groans of pain interpretation suggests). A true groans of pain signal (e.g., a scream when you stub your toe on a coffee table) is produced by an animal regardless of whether or not there is an audience listening in. At the other end of the communicative spectrum we find natural language, which is not only voluntarily produced almost exclusively for the benefit of an audience, but is predicated on assumptions by the speaker as to what the audience might or might not know about the subject being discussed. Not only do I need to have voluntary control of my language production when I am conversing with my wife around the dinner table, but I need to constantly be making guesses as to the contents of her mind in order to effectively convey useful information. For example, does she know that I saw *Midnight in Paris* on TV last night? If not, I should maybe tell her. But if she has not seen it yet herself, I had better not give away the ending. To produce a natural language, I am availing of a suite of socio-cognitive abilities culminating in theory of mind.

Evidence exists that alarm calls produced by chickens,[123] recruitment calls produced by pied babblers,[124] and threat displays produced by Siamese fighting fish[125] vary depending on the presence or composition of an audience. These audience effects are more widespread in the animal kingdom than once believed, and suggests that, whether or not these animals have conscious or voluntary control over their signals, their awareness of who's listening does influence signal production. For chimpanzees, it's more than just the presence of an audience that affects their alarm calling; they, like humans, might also be taking into account the contents of each other's minds. A 2011 study found that wild chimpanzees produce alarm calls associated with the presence of a dangerous predator (i.e., viper) more often when in the company of another chimpanzee who had either not seen the snake (i.e., had not focused visual attention on the snake), or who was too far away to have heard previous alarm calls. According to the authors, chimpanzees "monitor the information available to other chimpanzees and control vocal production to selectively inform them."[126]

There is no evidence that wild dolphins, like wild chimpanzees, produce signature whistles (or other referential/semantic calls) dependent upon the mental state of conspecifics. There is, however, evidence that dolphins might be able to attribute mental states to other agents, based on the studies involving gaze following and point comprehension/production. Similar abilities have been found in a number of species (e.g., goats, ravens). But dolphins, like chimpanzees and humans, might have a sophisticated understanding of the contents of other agents' attentional states (if not their minds), and can voluntarily produce communicative signals to communicate (representational) information. While this might fall short of full-blown theory of mind, it still bolsters their rating for Social Cognitive Aptitude. Unfortunately, this ability has not been found to apply to their natural communication systems as it does for chimpanzees, which is why it does not receive more than 3 on the Language Rating Scale.

Comparing Language Ratings for Five Species

Tallying up the ratings for dolphins on these ten ingredients reveals that they receive a total rating of 20 out of a possible 50. Humans receive a rating of 5 on all ten ingredients, yielding a Total Language Rating of 50. Regardless of where you come down on continuity, without a rating of (or very close to) 50, we cannot use the term *language* for whatever it is that dolphins are doing with their communication system. I will leave it to the reader to determine if the differences in ratings should be considered evidence of a functional discontinuity or not. For reference, we can compare the Language Rating of three other species (i.e., bees, chimpanzees, and chickens) to determine how unique dolphins might be with regards their linguistic skill. I've chosen chimpanzees because their cognitive skills are comparable to dolphins, bees because of their well-known language-like waggle dance, and chickens as a kind of baseline insofar as they are not typically considered language users.

It is obvious from the rating matrix that humans occupy a lonesome spot at the far end of the cognitive spectrum. Even when allowing chimpanzees high marks for evidence of an intentional gestural communication system, and an alarm call system where the caller takes into account the mental state and potential ignorance of the receiver, chimpanzee communication is a far cry from language. Despite (contentious) evidence for theory of mind for chimpanzees, and the fact that in experimental conditions they are adept at symbolically representing objects and events occurring in the past or future, we cannot escape the fact that chimpanzee communication and chimpanzee cognition in general lacks full-blown recursion, a discrete combinatorial system, or the capacity for limitless expression. The remarkable bee waggle dance is flexible enough to allow bees to transmit information concerning the direction of and distance to flower patches, which gives them high marks for displacement and mental time travel, as well as marks for the ability for limitless expression within the domain of

Language Rating Matrix Comparing Dolphins, Chickens, Chimpanzees, Bees, and Humans

Attribute	Dolphins	Chickens	Chimpanzees	Bees	Humans
Limitless Expression	0	0	0	1	5
Discrete Combinatorial System	1	0	1	1	5
Recursion	1	0	1	0	5
Special Memory for Words	1	0	2	1	5
Displacement and Time Travel	3	0	3	4	5
Environmental Input Required	3	0	1	0	5
Arbitrariness	4	4	4	3	5
Freedom from Emotion	2	1	2	3	5
Novelty Generation	2	0	2	2	5
Social Cognitive Aptitude	3	2	4	1	5
LANGUAGE RATING	20	7	20	16	50

"flower patch location" using a combinatorial system comprised of continuous variables (i.e., representations of distance/sun angle)—a system that is potentially even more powerful than one based on discrete variables.[127] But bee communication does not use recursion, nor take into account the intentional/mental states of conspecifics, and does not add new topics of discussion. And vitally, bee communication does not allow for limitless communication about topics other than flower patches and new hive sites. Chickens likely represent the norm as far as animal communication goes, and suggest that dolphins, chimpanzees, and bees possess abilities that are more language-like than your average animal. But even these stars of the animal communication world cannot fully bridge the gulf between ACS and language.

Some have argued that attempting to directly compare language to animal communication (as I am doing in my Language Rating Matrix) has little heuristic value,[128] that we simply don't know enough about either human language or animal communication to make meaningful comparisons,[129] and that all scholars can do is simply provide an opinion.[130] This is a valid point, although in this case I am not expressing my own opinion, but attempting to explain the majority opinion. After all, there must be good reasons why nearly the entire scientific community rejects the idea of dolphinese. However, I acknowledge that it's not the case that the general public agrees with the scientific consensus that the evidence for dolphin language is weak. When I speak to people about dolphinese, I typically receive a barrage of objections to the suggestion that dolphins do not possess a language. What follows is a rundown of the most common objections I've encountered, together with an explanation as to why they do not impact the overall conclusion that dolphinese is unlikely to exist.

Language in the Laboratory

Animal language programs—in which researchers teach an artificial symbolic communication system to an animal under experimental conditions—have revealed that dolphins are in an elite class of symbol-using species. Many of their accomplishments have led me to boost their Language Rating on a number of cognitive attributes, including arbitrariness, novelty generation, and memory for words. But proficiency in artificial language comprehension is by no means a direct test of how animals use their natural communication systems. Nonetheless, dolphins' performances on artificial language tests are often cited as "proof" that their own communication system functions like language. The Sirius Institute suggests that this is how they "know" dolphinese exists,[131] as do other dolphinese supporters:

> The lexical and semantic breadth of language the dolphins were able to understand (in captive studies) insures…the inherent ability of the dolphins to communicate using language.[132]

Thomas White argued a similar point in *In Defense of Dolphins* stating that "the idea that dolphins have advanced cognitive skills and something analogous to human language may turn out to be the simplest and most likely explanation for Phoenix and Akeakamai's performance in Lou Herman's experiment."[133] Herzing and White have elsewhere stated that Herman's findings are "highly suggestive of the fact that in their own lives they employ a capacity (the details of which have yet to be identified) that is equally as cognitively complex as the human capacity for language."[134] Lori Marino has echoed this sentiment, suggesting that part of the reason why dolphin brains are large is because they underpin a communication system that is "analogous to language," which explains their "prodigious artificial language abilities."[135]

I respectfully disagree with these conclusions. There is absolutely no question that, for example, the Kewalo Basin experiments showed that dolphins are capable of fairly complex abstract thinking and symbol comprehension, which is underpinned by some serious cognitive muscle that is rare in the animal kingdom. And as my Language Rating Matrix makes clear, many of these cognitive abilities are vital to the function of language. But it is pure speculation to suggest that the skill

Akeakamai showed in grasping abstract ideas like "through" or "yes," or that allowed her to understand that word order can change sentence meaning, is due to equivalent concepts existing in dolphin communication. This is a possible explanation, but it is not the most likely or the most parsimonious, nor is there any evidence to support this idea from the study of dolphin communication itself. Herman suggested that these skills might be latent cognitive abilities that bubbled up during the experiments, and which might never be realized in their natural communication systems.[136] It's also likely that novel ecological conditions (i.e., the foreign cognitive environment[137] of the experimental set-up) coaxed dolphins into using (and combining) cognitive skillsets that evolved to solve problems in domains unrelated to communication, let alone language.

This is the explanation for most species operating in a foreign cognitive environment, and there is no scientifically valid reason to make an exception for dolphins other than our own biases concerning the existence of dolphinese. Chimpanzees have also been shown to understand how word order changes meaning, to understand complex abstract concepts like "through" or "yes," and to report the presence or absence of objects. The explanations offered in the literature as to how chimpanzees accomplish these feats do not typically suggest that they are extrapolating and applying components of natural chimpanzee language (or the analog thereof) to artificial language experiments. Instead, the most widely cited and likely explanation is that these cognitive skills are used in domains other than language or communication, but can be co-opted to meet the demands of novel experimental situations. As W. Tecumseh Fitch described it, "the capacity of an organism to master some human-invented system reflects their cognitive abilities, rather than those underlying their own species-typical communication system."[138]

If we want to determine whether or not a dolphin's own communication system utilizes the cognitive skills we observe in artificial language programs, we need to test directly for them, and not infer them from the results of peripherally related experiments. Inferring the existence of wild/natural behavior from an animal's performance in a controlled experimental test of cognition can be the wrong approach. In a famous experiment, capuchin monkeys were taught to use tokens to "pay" for food rewards.[139] As the "price" of food items went down, the monkeys reacted the way any human would—they bought more of it. These experiments—and others that followed—showed that monkeys understand some basic principles necessary for commerce (e.g., reference dependence, loss-aversion). Does this experiment allow us to conclude that the capacity for loss-aversion evolved in the wild as part of a capuchin monkey social system wherein currency is used to purchase food commodities, or something analogous to it? No. It reveals that innate cognitive biases like loss-aversion can be co-opted by the demands of a commerce experiment, but these biases almost certainly evolved to address a disparate ecological scenario. It would be very strange indeed to suggest that capuchin monkeys must have something analogous to a commerce system in the wild as being the simplest explanation as to their performance on this particular experiment.

The discovery of cognitive attributes known to underpin language in the lab will, at best, alert us to the possibility that an animal's natural communication system might make use of these skills. But even when functional analogs have been found in both artificial language systems and dolphin communication (e.g., object labeling and signature whistles), these behaviors might be supported by very different cognitive attributes. After all, dogs do not produce referential signals like alarm calls, and yet Chaser the Border Collie has learned more labels than any animal on the planet. Any link cited between artificial language programs and the structure or function of ACS should be considered something akin to either an untested (but testable) hypothesis or wishful thinking until such time as evidence for its existence in ACS is found. As we have seen, this evidence is currently lacking, which is why Louis Herman has concluded that, despite dolphins' symbol manipulation skills in the lab, "there is no evidence that dolphins in the wild have anything approaching a human natural language."[140]

Dolphin Language Production

Dolphin researchers have occasionally been criticized for failing to test for dolphins' ability to produce (as opposed to comprehend) symbols and signs. The research at Kewalo Basin involving Ake and Phoenix was specifically designed to test only comprehension, which was a direct result of the criticism that other animal language programs had been receiving concerning the lack of control over the experimental context and stimuli seen when animal subjects were engaging in simultaneous symbol production and comprehension. Although it is more difficult in an aquatic environment to create a symbol system that dolphins could use for artificial language production, it is not correct that this approach has been ignored by the scientific community. In fact, before, during, and after the heyday of the Kewalo Basin experiments, researchers have been testing for dolphins' ability and interest in producing meaningful utterances using language or artificial symbol systems. Just recently, an experiment involving a beluga whale (not a dolphin species, but a toothed whale relative) called Nack showed that he was able to both comprehend and produce four acoustic calls that corresponded with four objects.[141] The research involving artificial language production for dolphins has seen similar results, but never managed to move beyond them.

In a now iconic experiment from the 1960s, John Lilly and his research assistant Margret Howe attempted to teach Peter the dolphin to not only understand, but "speak" English.[142] Margret lived side by side with Peter in a semi-flooded two-room house for two and a half months. Results suggested that Peter was able to passably mimic a couple of human vocal sounds, but the overall experimental design left much to be desired, and provided little evidence of language production. In the early 1980s, Lilly embarked on a research program called Project JANUS (Joint Analog Numerical Understanding System) at Marine World in California, with the intention of developing a shared two-way communication human/dolphin code. Using computers that transmitted and received sounds to the dolphins underwater, the goal was to establish a "human/dolphin dictionary" of whistle symbols associated with objects and actions.[143] Although Lilly promised a breakthrough in human-dolphin communication within five years with JANUS, no results from this project were ever published.

A long-term research program (occurring at Marine World at the same time as Project JANUS) developed by Diana Reiss in the 1980s involved dolphins learning to associate symbols on an underwater keyboard with artificial whistles that referred to objects in their environment (e.g., ball, fish).[144] The dolphins were able to make these associations and to mimic the artificial whistle symbols. A single anecdotal account suggests that one dolphin might have produced the "ball" whistle as a request (i.e., as a referential label) when communicating with another dolphin. Other than that occasion, the dolphins mostly produced the whistles when interacting with the objects in behaviorally appropriate contexts (e.g., producing the "ball" whistle when playing with the ball). A later study by Reiss and colleagues revealed that dolphins are able to spontaneously learn whistles that were generated by experimenters during interaction with specific objects, although they did not produce the appropriate whistle when playing with associated object as they did in the first study.[145] The main finding of these studies (aside from the spontaneous generation of a novel whistle as described earlier) is the ease with which dolphins (unlike most mammals) can imitate whistles. Similar studies have been conducted[146,147] showing that dolphins are able to learn artificial whistles that could then hypothetically be used in studies of artificial language production (and thus two-way communication), although the two-way experimental stages were not reported on (or never took place).

Ken Marten, working at Earthtrust's Project Delphis lab in Hawaii, together with Fabienne Delfour, developed an infrared underwater touch-screen in the early 1990s, allowing the dolphins to trigger and manipulate video and audio. Marten and Delfour later began a project to co-create a shared symbolic whistle system that involved associating word meaning with various discrete whistle sounds produced by the dolphins.[148] Marten did not provide food rewards to his dolphin subjects, but used the dolphins' natural curiosity to motivate them to participate in his experiments. Sadly, Marten died in 2010, with no breakthrough in establishing dolphin production competence having been reported in the scientific literature.

In the early 1990s, researchers at the Living Seas in Epcot used a keyboard system whereby a diver or a dolphin could trigger a key to indicate a location in their tank containing an object of interest.[149] The goal was to entice the dolphins to use the keyboard to refer to these areas/objects, which they eventually did after six months of training. This experiment was phased out in favor of the more successful pointing experiments that arose from this work.

Denise Herzing began a long-term, multi-phase experiment starting in 1997 to develop a two-way communication scenario whereby wild dolphins would learn to use an artificial symbol system to spontaneously share information with human researchers.[150,151] The initial phase involved researchers interacting with well-researched groups of wild spotted and bottlenose dolphins off the Bahamas, and producing a series of tones from a wristband computer while engaging in specific behaviors (e.g., dive, play keep away). Later, a keyboard with a combination of audio/visual symbols was introduced, with each symbol associated with four objects/concepts: scarf, rope, sargassum, and bow ride. In order to get the dolphin interested in using the system for communication, Herzing's team based their research on the social-rivalry framework that Irene Pepperberg had used to train Alex (and her other parrots) to use symbols. This involved the researchers modeling the desired behavior (e.g., getting another human's attention and then requesting an object by pointing or via the symbol system). The only reward the dolphins received would be the fun they had in playing with the objects in question (e.g., playing keep-away) together with each other or the human researchers, an activity that they had engaged in regularly before the experiments began.

In the four years that these experiments were conducted in the Bahamas, the dolphins responded well to (and even spontaneously produced) pointing behavior at play objects, and seemed to be quite interested in interacting with the researchers in general. Some dolphins did orient to specific keys on the keyboard as a possible request for an object/action, and on one occasion responded to a request initiated by a researcher to fetch a play object. They did not, however, appear to imitate or produce the artificial whistle sounds, and never actually touched the symbol keys (which is not something a wild dolphin would likely do). This research is ongoing, with a new phase of the project incorporating updated technology capable of both generating audio symbols and recording dolphin vocalizations (and hopefully imitations of these symbols) underwater.[152]

It's probably fair to say that the results of language production experiments for dolphins have been less impressive than for species like chimpanzees, gorillas, bonobos, and parrots. Contrast if you will the communication board used by Kanzi the bonobo, which contains hundreds of symbols (lexigrams) that he uses to represent objects, actions, and abstract concepts, with the communication devices used by dolphins, who have only ever used/produced a handful of symbols (whistles or lexigrams) to represent objects. Surely part of the reason for dolphins' poor performance lies in the difficulties in conducting these sorts of experiments in an aquatic environment. Considering dolphins perform similarly to bonobos and chimpanzees on tests of symbol comprehension, it would be expected that their production would be similarly impressive. Nonetheless, dolphins have been given the tools to establish two-way communication with human beings using a variety of artificial symbol systems and under numerous experimental and wild conditions, and the fact that they generally seem less interested than their primate (or avian) cousins in using these systems is an interesting finding. Wild dolphins' comprehension and use of pointing-like gestures in the context of play in Herzing's experiments is an interesting finding insofar as pointing seems such an unlikely skill for an untrained dolphin to understand/use, and one vital to establishing two-way communication, but their behavioral reactions in these experiments vis-à-vis the objects and symbols more closely resembles that of domestic dogs[153,154] than the kind of symbol production seen in the great apes. My personal interpretation of this line of research is that dolphins show a clear, rather extraordinary desire to interact socially with human beings, but not necessarily to communicate with them via symbol systems. This could be taken as a sign that the fundamental aspects of language-like communication needed to succeed in these types of conditions are utterly foreign to them. Or it could be a sign that scientists have not yet hit on the correct motivational key to success. Herzing's social-rivalry play framework seems a likely candidate for success, on paper at least,[155] but only time will tell if it results in the dolphin-symbol-production breakthrough that has been long promised, but never realized. Regardless, production of symbols with intent to communicate is clearly not something dolphins take to easily, despite years of testing.

The Building Blocks Fallacy

Some have argued that the variety of sounds produced by dolphins—all those clicks and whistles—could easily be combined by a discrete combinatorial grammar system to produce an unlimited number of "words" that give dolphin communication the property of limitless expression.[156] This is true. In fact, it is possible to take any elements of the natural world and use them as the basis for a discrete combinatorial system. A snapping shrimp could snap its claws in specific patterns—like Morse code—to convey unlimited information, assuming it had the cognitive wherewithal to map specific meaning onto discrete snap-patterns. There is nothing unique about the concept of a phoneme produced in the human vocal tract that is essential for natural language. Indeed, human sign languages make use of hand gestures and facial expressions as phonemes in their combinatorial systems, yielding a system of communication that is unequivocally considered a natural language by the definition I've provided. Evolution could have used hand claps, blinking, or foot stomps as the phonemes upon which to build the natural languages used by humans. There is no reason why chimpanzee gestures, cricket chirps, or dolphin whistles couldn't also have been co-opted by evolution as the building blocks of a discrete combinatorial system with all the properties of natural language. But, and this is the crux of the issue, there is no evidence that they have. No species other than *Homo sapiens* has evolved the necessary cognitive muscle to actually combine potential building blocks into a system as open ended as natural language. The potential for combining discrete elements to create limitless expression exists everywhere in nature and within the communication systems of all species, including dolphins. But potential does not equal existence. An appeal to the existence of potential building blocks as proof of language in dolphins is logically untenable, and is known as the Building Blocks Fallacy.

Information Theory

An argument in support of the idea that dolphins have a language-like communication system is that the use of information theory to study dolphin vocalizations has revealed that dolphin communication is extraordinarily complex and that the structures we find in their natural communication system closely resemble what we find in human languages.[157] The field of information theory was developed by mathematician and cryptologist Claude Shannon.[158] In his landmark article from 1948, Shannon showed how communication systems can be described as containing bits (a term he coined) of data. The average amount of information (i.e., number of bits) that a symbol contains is what Shannon called a measure of its entropy, which is now referred to as *Shannon entropy*. In the 1960s, John J. Dreher calculated the Shannon entropy contained in his dolphin whistle data set, and reported that the number of bits for each whistle contour ranged from 1.60 to 2.17, which was similar to what is found when you calculate the entropy of letters occurring in written English (i.e., 2.02).[159] The possibility then is that dolphin whistles might function like human words in terms of the kind of information they could potentially transmit. The stage had now been set for using information theory to calculate the potential informational content—and possibly even complexity—of dolphin vocalizations. However, despite support from prominent cetologists for this line of research,[160] information theory would not be used again (in the United States) to study dolphin vocalizations for another forty years.[161]

Soviet scientists applied information theory to the study of bottlenose dolphin vocalizations in the 1980s.[162] Instead of calculating Shannon entropy, Vladimir Markov[163] and his co-author, Vera Ostrovskaya looked at the rank distribution of vocalization types, to see how closely this distribution matched that found in natural languages. By plotting the frequency that individual language units (e.g., letters, words) occur in a given sample of language (e.g., block of text, speech transcription), mathematicians have found a strange relationship between a unit's frequency and its rank on the frequency scale. For example, the most frequently occurring word in an English text sample might (hypothetically) be "the," which we see occurring 1,000 times. It ranks as the #1 most frequently occurring word. The #2 most frequently occurring word is "of," which occurs about 500 times. Note that "the" occurs twice as often as "of." The #3 most frequently occurring word, "at," occurs approximately 250 times—exactly half as much as "of." This relationship between frequency and rank is not just hypothetical—it is a real phenomenon that is found in nearly every human language, and is referred to as Zipf's law (or Zipf's distribution), named for Harvard linguist George Zipf. A graph plot of this distribution for most languages results in a straight line (inverse relationship) slope of −1 (i.e., Zipf's statistic). The reason languages follow this distribution and generate Zipf's statistic so doggedly is unknown. When Markov and Ostrovskaya ranked the frequency of different vocalization types from their dolphin vocalization data set, they found something akin to Zipf's statistic, which led them to conclude that the dolphins were using optimal encoding which is nearly identical to that observed in human speech.

The revival of information theory in the study of dolphin communication in the West was realized by Brenda McCowan, a behavioral biologist at University of California, Davis, who teamed up with, among others, Laurance R. Doyle at SETI. Doyle, an "alien hunter" like his colleague Seth Shostack, applies information theory to find intelligence in both the communicating signals of animals, and any signals we might detect emanating from alien life. In a series of articles, these researchers used various applications of information theory to argue that Zipf's law and Shannon entropy is a useful means of looking at structure and complexity in animal communication signals. Similar to the results obtained by Markov and Ostrovskaya, a Zipf's statistic of −0.95 was found for McCowan's dolphin whistle data, which closely matched what is found in samples of human languages.[164]

McCowan and colleagues argued that Zipf's statistic can be used as an indicator for potential information capacity of an animal's vocal communication system, with a Zipf's statistic approaching −1 (i.e., more like that seen in human languages) as likely containing more structural diversity in the vocal repertoire. Zipf's statistic might then be showing that dolphin communication involves vocalizations that contain an optimal balance between signals carrying vital information, and extra "redundant" signals that protect or buffer the system from loss of information in a noisy communication channel. It has been proposed that it is this "optimal" balance between bits that carry information and bits that do not carry information that Zipf's statistic is actually reflecting in human language. Markov and Ostrovskaya also reportedly found this in their analysis of dolphin whistles, which they referred to as "optimal encoding."

The analysis of the amount of complexity (as opposed to structure or diversity) within this communication system can be done by calculating higher-order (Shannon) entropy, for which these authors provided only a preliminary analysis due to lack of data. Their entropy analysis allowed them to calculate the probability that one whistle might follow another, which could be used as an indication that sequential whistle production is governed by a non-random (i.e., structured) and complex communication system—something akin to syntax. The authors argue that more structure of this type equates with a higher potential "communication capacity," which is something that could be calculated using their technique for producing an "entropic slope."

McCowan and her co-authors argue that by calculating structural diversity (with Zipf's statistic) and structural complexity (with entropic slope) we can 1) study how individual species learn to use their communication system (by noting the differences in repertoire complexity between young and mature individuals), 2) formulate predictions of signal structure and complexity based on what is known of species' ecological and evolutionary history, and 3) directly compare the potential complexity of two different species' communication systems. Critics of McCowan and colleagues' use of information theory in the study of animal communication point out that Zipf's statistic tells us nothing at all about whether or not dolphin whistles contain any communication content, let alone if this content is language-like in terms of its semantic or syntactic value.[165] But McCowan and colleagues defend their approach by stating that while it is true that information theory cannot tell us if dolphin communication is equivalent to language, or that it contains any information or structure at all (and certainly not meaning in the linguistic sense), when we do find a Zipf's statistic that is close to −1, it means that dolphin communication *might* contain structural complexity, and probably deserves further study.[166,167] In other words, information theory is simply a measure of potential signal diversity and potential signal complexity, not language. To find the presence of information, meaning, and language-like structure in animal signals requires boots on the ground—actually observing animals as they produce their vocalization within appropriate behavioral contexts, and using the most powerful tool available for detecting patterns and meaning in animal vocalizations: the human brain.[168]

3D Holographic Communication

In the book *Communication Between Man and Dolphin*, John Lilly once wrote that studies of echolocation in dolphins "lead to the conclusion that the basis for the postulated language of dolphins, 'dolphinese,' is based upon the construction by central processing of 'acoustic pictures,' which are the basic elements of the postulated language."[169] These thoughts were published in 1978, and I suspect that the many experiments in the 1960s and 1970s suggesting that dolphins do not likely communicate complex arbitrary information via their vocalizations led Lilly to abandon his original ideas of how dolphinese might function in favor of this new concept. No longer was the idea that dolphins combined discrete symbolic vocal-units like a natural language, but instead they were transmitting holographic 3D sound images using their biosonar. This remains a popular idea, with Global Heart, Inc. dedicating their own line of research to deciphering the holographic "sono-pictures" that dolphins are hypothetically sending to each other via their echolocation clicks.[170]

There remains, however, no evidence that this is something dolphins do. Global Heart, Inc.'s research has not yet been published in the peer-reviewed literature, so it remains impossible to evaluate, or to know if they are basing their ideas on empirical evidence stemming from sound methodological techniques and data analysis, or simple conjecture flowing from Lilly's original ideas. In fact, the mechanisms underpinning dolphin echolocation are, thanks to the US Navy's intense interest in this area, perhaps the most studied aspect of dolphin behavior.[171] And what we know of how this process works (which is quite a bit at this point) suggests that dolphins cannot and do not alter parameters of their outgoing clicks in order to reproduce and transmit anything akin to sonic images.

The basic idea of echolocation is that a dolphin produces clicks that bounce off objects in the underwater environment producing a click echo. The dolphin listens for the echo, and, based on the information contained in the structure of the echo, is able to discern the location, size, movement, internal structure, etc. of that object. When the sound waves from a single dolphin click come in contact with an object, like a fish, a number of things will happen. A fish is comprised of a variety of organic material that has different reflective properties. The swim bladder (filled with air), for example, is highly reflective whereas soft tissues, like eyes, are far less reflective. As sound waves from the click penetrate the fish's body, these different tissues will create different echoes with different properties, producing what acousticians call backscatter. The swim bladder might produce an echo that, for example, is quite loud, whereas the liver absorbs the sound waves and produces a very quiet echo. As the sound waves penetrate the fish's body, a series of echoes will begin to be reflected at different times, so that the echoes from the head, for example, arrive back at the dolphin before the echoes from the fish's tail. There will also be differences in the frequency spectrum of these backscatter echoes, with some echoes containing more or less energy at various frequencies. Thus, the series of echoes that return to the dolphin from a single click are quite messy, but undoubtedly contain a lot of useful information (for a dolphin). Scientists have been particularly interested in how this process works as it has obvious implications for the design of our own sonar systems. Researchers have been able to use artificial click echo generators, which are used to create and tweak click echoes that are then transmitted to a dolphin, to determine what properties of click echoes are vital for dolphins in discrimination tasks.[172] In other words, scientists have a very good idea at this stage as to how click echoes differ in structure depending on the object being ensonified, and what aspects of these echoes dolphins are using to extract information.[173]

Scientists have also investigated the structure of the outgoing clicks produced by dolphins. Dolphins change the parameters of their outgoing clicks in a number of ways for a number of different reasons.[174] They can increase the intensity (volume) of the clicks in order to, for example, receive echoes over longer distances, penetrate clicks deeper into an object, or to compensate for noise in the environment. They will also alter the frequency (spectral) makeup of the clicks depending on the requirements of the discrimination task.[175] They can steer the direction and shape of their echolocation beam by changing the parameters of their clicks, the shape of their melon (i.e., the fatty deposits in their forehead), and by moving the position of their head (e.g., head scanning).[176] Numerous studies suggest that, in general, dolphins change the parameters of their outgoing clicks for the sole purpose of maximizing the information they will be receiving in the click echoes.[177]

What this all means in terms of the 3D holographic communication hypothesis is that there is absolutely no evidence to support the idea that outgoing echolocation clicks are being manipulated by a dolphin in order to transmit sonic (3D) images. In order for a dolphin to transmit the kind of detailed object information we know is found in click echoes, dolphins would need to generate an outgoing click that bears some sort of resemblance to click echoes and the complex backscatter generated by an object. This is most clearly not the case. Outgoing clicks are fundamentally different in structure to click echoes. Imagine if you will the kind of complex backscatter that would be produced by echolocation reflecting off a large school of fish. Dolphins are able to sort through this acoustic mess to pull out information that is useful to them.[178] But it's obvious, based on what we know of how dolphins produce outgoing clicks, that this complex backscatter nightmare is not anything that a dolphin itself could generate. Consequently, the idea that a dolphin could reproduce such a scene as a "sonic image" using its outgoing clicks does not seem in any way plausible. While it has been shown that dolphins can discern object information by listening to each other's click echoes,[179,180] there is no evidence that they do—or even could—transmit similar information directly to each other with their outgoing echolocation.

The Dolphinese Holdouts

The discussion above provides an overview of why mainstream scientific thinking has come to the conclusion that dolphins do not have a language. We have, however, heard arguments from a handful of people—both scientists and non-scientists—who disagree with the general consensus. The *dolphinese holdouts* typically fall into one of three categories: 1) true believers, 2) language redefiners, and 3) code crackers. An analysis of their objections and arguments follows.

The true believers

I've exchanged emails with a number of very thoughtful people who've listened to my reasoning as to why scientists don't think dolphin communication functions like human language, and their final reaction is usually something along these lines:

> "What I'd really like to know is, what evidence is there that dolphins do not possess an evolved faculty for language?"
>
> "It does not appear, however, that it has been definitively shown that the 'language instinct' is absent from the dolphin."[181]

These seem like very clear and insightful gotcha questions. But they are actually questions that are the result of a common (albeit sinister) logical fallacy: *argumentum ad ignorantiam*, otherwise known as the appeal to ignorance. To explain the pitfalls with this line of reasoning, I must invoke the wisdom of Bertrand Russell, whose celestial teapot analogy brings the problem into sharper focus:

> If I were to suggest that between the Earth and Mars there is a china teapot revolving about the sun in an elliptical orbit, nobody would be able to disprove my assertion provided I were careful to add that the teapot is too small to be revealed even by our most powerful telescopes. But if I were to go on to say that, since my assertion cannot be disproved, it is intolerable presumption on the part of human reason to doubt it, I should rightly be thought to be talking nonsense.[182]

The false argument being offered here is that because science has not yet or is unable to prove that dolphins do not have a communication system that functions like human language, we really should assume that they do. But as Russell points out, this is not how it works in the real world of empirical inquiry. The burden of proof is always on the person making a claim about the existence of a phenomenon—in this case, dolphinese. There is also no evidence that dolphins cannot time travel, cannot bend spoons with their minds, and cannot shoot lasers out of their blowholes, but it would be strange indeed to insist that they can do these things simply because we haven't yet proven that they can't.

This argument is also built on the idea that science has not been testing—or perhaps not adequately testing—for the presence of language in dolphins. But this is patently false. Scientists have been testing the hypothesis that "dolphins have a language" for decades, breaking it into component parts, and searching for these ingredients in the natural communication system of dolphins, or via experimental tests of communication and cognition. The results—described in detail in this section—suggest there is no evidence for full-blown language. Science has collected enough evidence of absence for the existence of dolphinese to have led to the conclusion that it does not likely exist. To continue to maintain that it does exist will require compelling evidence in support of the hypothesis, not an appeal to logical fallacies. The existence of dolphin language—in the way I am defining language—would be a discovery that fundamentally challenges our ideas of how animals communicate. It is an extraordinary claim, which is precisely why it needs extraordinary evidence to support it. Imploring that "you haven't proven that it doesn't exist" seems a hollow plea in comparison. Unlike the subjective experience of emotions in animals (which is ultimately impossible to test for via the scientific method), or Russell's teapot, language properties in dolphin communication are something we could easily find in ACS; and the lack of evidence for the existence of these properties is what makes it logically tenable to assume their absence.

A true believer, however, is not swayed by this argument. For those enamored of the idea of full-blown language skills in dolphins, there is not likely to ever be enough evidence of absence to inspire them to change their mind. Lilly's writings, and the cultural reference points that they incited, are too deeply embedded to be extracted so easily. I offer this final thought to the true believers (who I suspect constitute a very small percentage of the *dolphinese holdouts* to begin with): even if science has been going about this all wrong, and dolphins do have a language every bit as complex as human language that simply does not rely on the ingredients I've identified as necessary for language, they absolutely must satisfy the criterion of limitless expression if we are to consider whatever it is they're doing as language-like. Feel free to invoke all the hypotheses you would like as to how dolphins manage to accomplish this without the use of the other language ingredients I've listed (e.g., 3D holographic communication, psychic connections), but until there is evidence that dolphins are exchanging semantically rich signals on unlimited topics (an infinitely testable—and already tested—hypothesis), these ideas remain nothing more than science fiction. All it takes is one well-designed experiment to show that dolphins can and do transmit semantically rich signals to each other and the whole scientific community will be happy to change their opinion on the topic of dolphin language. After all, scientists are open-minded when it comes to being swayed by evidence—a trait not usually associated with a true believer.

The language redefiners

The language redefiners are a fairly reasonable group of objectors who are simply uncomfortable with the idea that language should have the kind of narrow definition that I've provided. This is often a response to the (outdated) idea that animal scientists are keen to uphold the uniqueness of humans and human behavior by designating ACS as a lesser version of language. By their reasoning, we should adopt a more inclusive definition of language to stop perpetuating the myth that there is a fundamental difference between human and animal minds.[183] I am sympathetic to this argument, but I do not consider it very helpful. We could move the goal posts a bit, and suggest that animals like dolphins or chimpanzees—which show skill in a number of cognitive areas necessary for language—have an ACS we should label language. But this is an exercise in semantics (or possibly politics), which detracts from the nature of the problem. There are both quantitative and qualitative differences in the cognitive competencies of humans and dolphins, and whether you prefer to term these variances a functional discontinuity or not does not change the fact that they exist. Trying to uncover the nature of the differences is a helpful approach to understanding both animal communication and human language—regardless of how you choose to label them, or how far apart you portray them on a continuum of behavioral complexity. The bottom line is that dolphins do not perform as well as humans on any of the cognitive tests traditionally associated with language-like behavior, and it's well worth asking why not.

The code crackers

There is a lot about dolphin communication that we do not know. It is difficult to record dolphin vocalizations, or observe their behavior in the wild. We are limited by recording technology and the challenges of field work in an inhospitable (to humans) environment. Despite these challenges, dolphins have received much research attention in the past fifty years, especially in the areas of cognition and communication, with long-term research projects (like those conducted by Richard Connor, Kathleen Dudzinski, John K. Ford, Denise Herzing, Janet Mann, Randall Wells) rivaling the richness of studies conducted on terrestrial species. Consequently, although we've much to learn about their communication system, it's not entirely unreasonable to make educated (i.e., scientifically informed) guesses as to the structure and function of dolphin communication. So, despite the fact there is much we do not know, we're beginning to understand what dolphin communication is all about.[184]

Unfortunately, the ever-present scientific caveat that "there is much we do not know" has allowed dolphinese proponents to slip the idea of dolphin language in the back door. If we don't yet understand what all these vocalizations are used for, it is argued, then there might still be an underlying language system that we've just not yet been able to decipher. We just haven't cracked the code, plead the code crackers. But why should we think this? What is it about dolphin vocalizations that lead anyone to suspect that there is a language code present, where it's not present in chickadee songs or howler monkey hoots? The answer likely lies at the feet of John Lilly, and not the corpus of research on dolphin communication itself.

The idea of "cracking the code" has been around ever since Lilly first proposed the idea of dolphin language. Half a century later, we are no closer to cracking this supposed code, as Lilly had predicted. Dolphin communication is certainly intriguing—and dolphins do have a large vocal repertoire, as Lilly first pointed out (although nowhere near as large as some bird species). But the resemblance this bears to language is superficial, as it is for other animal communication systems. The evidence I have outlined in this section suggests that dolphin communication is no more language-like than any other ACS, which makes it strange to continue to single out dolphins as possessing a language code.

But it's not just the diversity of dolphin vocalizations that the code crackers cite as reason to believe in dolphin language, it's the behavioral complexity of dolphins themselves. But this too is a strange argument. Chimpanzees, which are the most human-like of all animal species, possess many of the complex socio-cognitive skills we might except of a language-using species. If we had to place bets as to which species possess a language based on social structure, cognitive abilities, or ability to manipulate symbols in an artificial language experiment, then chimpanzees would be the horse to pick. But these behavioral criteria are clearly not useful indicators of an animal's ability to produce language-like skills with its own communication system. Prairie dogs, which don't hold a candle to chimpanzees or dolphins in a battle of wits, possess one of the most complicated, semantically rich signaling systems on the planet. And honeybees, which are *insects* for Pete's sake, satisfy many of the criteria of language with their waggle dance. Clearly the relationship between traditional ideas of what makes good language-user candidates (i.e., which species act the most like humans) are out of step with the reality of which animals in fact display language-like skill. If anything, the strongest candidates so far for having a language-like system are rodents, so it would be more logical to speak about cracking the prairie dog code than the dolphin code.

Code crackers also appeal to the findings of potential structure in dolphin vocalizations via information theory as evidence of an underlying language code. But we've seen that information theory is not necessarily suggesting that dolphin communication is any more language-like than other ACS, and has little to tell us about language-like structures in dolphin communication.

It is not an entirely unreasonable or an unscientific approach to suggest that we cannot rule out the existence of unknown complex language-like structure in dolphin communication. One does not necessarily need to produce a Russell's teapot argument to make a case for this as a code cracker. When we combine the results of studies of dolphins, we find that they are intelligent, behaviorally complex animals with a diverse vocal repertoire and unusual (compared to most animals) skill at comprehending artificial symbol systems. But as I hope I've been able to show, each of these facts on their own—or even when taken together—is not a sufficient argument to make dolphin language any more probable than orangutan language or dog language. The gap between human language and ACS as a whole—this great functional discontinuity—makes it very difficult to insist that there is a code there to crack. As the years go by and we learn more about how dolphin communication actually works, and without new evidence of language-like properties in dolphin communication, the code cracker argument will start sounding more and more like the protests of a true believer.

Retiring Dolphinese

Is comparing animal communication systems to language even a valid approach to begin with? It is arguably unjustifiable from an evolutionary standpoint to view ACS as a collection of systems that are not quite fully formed versions of language. This assumes language itself is some end point on the conveyor belt of evolutionary progress. But this teleological view of the situation constitutes bad science, and confuses the issue. Language is an exquisite behavior forged by evolution—a trait seen in humans and no other animals. We can ask questions as to how and why it evolved from an evolutionary standpoint in the same way we can ask how and why dolphin communication evolved, without suggesting that one system is a lesser or greater version of the other. Each system does the job that evolution intended it to do. In the same vein, a dolphin's blowhole is an exquisite morphological design, but it's fairly unproductive to compare blowholes to human nostrils if we're looking to gain a deep biological understanding of the form, function, and evolution of blowholes. A blowhole must be understood for what it is, and not what it is in comparison to something else.

That being said, compiling a Language Rating score to see how well dolphin communication compares to natural language is a helpful exercise. It makes us think about what is going on in the human brain to create language in the first place, and teaches us about the dolphin mind by teasing out the aspects of language behavior that they exhibit. The end result, however, sounds like a scorecard showing the inadequacy of dolphin communication: they don't achieve limitless expression, they don't quite satisfy the criteria for full-blown socio-cognitive aptitude, they don't have a discrete combinatorial system, etc. If we become obsessed with searching for language ingredients in dolphin communication, we will continue to butt heads with all this negativity, and lose sight of what's important. By focusing on the search for phonemes, or language-like syntax, which, according to everything we know about ACS, are extremely unlikely to be present, we might miss the presence of real phenomena that, while not language-like, are nonetheless important to dolphins. The search for dolphinese is a distraction—a red herring that saps our limited resources, and funnels budding scholars away from the important problems in the field that require our full attention. It has not been Lilly's ideas or research questions that have provided breakthroughs in the study of the dolphin mind or dolphin communication in the past few decades—it has been the hard work of scientists who've long since abandoned his dolphin-language-deciphering goals. It is time now to fully retire dolphinese, and embark on a more fruitful campaign to unravel the real mysteries of dolphin communication.

6 A Most Gentle Mammal

Everyone loves the king of the sea, ever so kind and gentle is he.[1]

Lilly's idea of dolphin intelligence included the notion that dolphins had a sophisticated ability to control their emotions,[2] which explained why they refrained from lashing out at him when he inserted electrodes into the brains of unanesthetized dolphins. This idea has blossomed into the popular idea that dolphins are unusually peaceful, friendly animals that live in harmony with other creatures in their environment. The New Age book *In the Presence of High Beings: What Dolphins Want You to Know* focuses a lot of attention on this notion. The book lists "unfailing friendship and kindness" as the number one "gift" or "special trait" that dolphins desire to share with humankind:

> The quality I first noticed in the dolphins was how consistently they offer loving friendship and kindness to humankind. In addition to harmony in their environments, among themselves, and in music, dolphins enjoy harmony with other species.[3]

Piggybacking on this idea is the assertion that dolphins are socially complex animals. Having "rich social lives" forms part of the argument for personhood in dolphins,[4] and the sophisticated social structure displayed by killer whales formed part of the justification for PETA's lawsuit naming killer whales as plaintiffs in their anti-slavery lawsuit against SeaWorld:

> Orcas live in large complex groups with highly differentiated relationships that include long-term bonds, higher-order alliances, and cooperative networks. They form complex societies with dynamic social roles in intricate networks, many with distinctive cultural attributes in vocal, social, feeding, and play behavior.[5]

So what does the science have to say about the idea that the intelligent dolphin mind has rendered them unusually social, friendly, and peaceful animals that live in harmony with their environment?

Peaceful Behavior

There certainly is no shortage of both anecdotal and scientific evidence of friendly and peaceful dolphin behavior. Behaviors like supporting sick, drowning, or dead conspecifics (and humans) have been part of the Western cultural narrative since antiquity. If one is lucky enough to have observed dolphins in the wild, especially underwater, it is very likely you will have witnessed dolphins being playful and affectionate with one another, both among and between different dolphin species.[6] Dolphins are often in physical contact, using their pectoral fins to gently rub each other's bodies, or otherwise maintain contact.[7,8,9] Although often wary and shy when it comes to encountering humans in the open ocean, there are hundreds of cases of wild dolphins seeking out human companionship, including obvious signs of friendly curiosity and even close physical contact.[10]

Of course, friendly or affiliative social behavior is not something unique to dolphins. Almost any animal species—from slugs[11] to chimpanzees[12]—engages in social behaviors involving gentle, peaceful, tactile exchanges or comforting/soothing/consolation behavior when a close social partner or family member is in distress. And, like dolphins, friendly interspecies social encounters have been witnessed in species ranging from monkeys[13] to birds.[14] But it's not the presence of affiliative behavior that will make or break the case that dolphins are unusually friendly or live in peaceful harmony with each other and their environment, rather it would be the *absence* of aggressive behavior.

It's unlikely these days that even the staunchest proponent of the idea that dolphins are peaceful creatures is unaware that dolphins sometimes display violent behavior. With National Geographic's documentary *Dolphins: The Dark Side*[15] suggesting that "loveable-looking 'Flipper' is also cunning, aggressive, and brutal,"[16] the general public has been exposed to ample non-peaceful dolphin behavior. But the question remains as to whether this violent behavior is either 1) common, or 2) abnormal for dolphins. *Dolphins and Their Power to Heal* dismisses aggressive behavior as something that arises when dolphins are "provoked,"[17] and contends that abnormal situations like captivity are what create "social stresses and pressures [that] will make the normally placid, compassionate and good-natured dolphin behave in an uncharacteristically selfish and aggressive way."[18] Other authors have reached similar conclusions, suggesting that any aggressive behavior witnessed in dolphins is a product of abnormal (usually human) stressors or artificial environments.[19] Captivity is, according to some, the primary cause of aggression observed in dolphins:

> The most prominent changes seen in semi-tame and captive dolphins include the onset of competition for food and subsequent fighting and aggression among them, even resulting in wounds. Yet I have never witnessed a single altercation in the wild and only once in ten years have I seen a free dolphin with open wounds.[20]

The Sirius Institute claims that "from what we know, dolphins treat people with kindness except under extreme provocation."[21] *Provocation* is the buzzword often used to explain violent behavior observed in dolphins, whether in captivity or in the wild. The general idea is then that aggressive behavior is abnormal for dolphins, with the implicit notion that aggression itself is an undesirable anomaly as opposed to a valuable and normal class of behaviors common to dolphins, humans, and other complex social mammals. But, as we shall see, ample evidence of "unprovoked" aggressive behavior in wild dolphins makes the idea that "aggression is abnormal for dolphins" a rather hard pill to swallow.

Porpicide and infanticide

I will begin with a discussion of fatal interspecies aggression with a side order of infanticide. Frankly, this is probably the only example you really need in order to scratch dolphins off of your nice list. As some overly dramatic news headlines have revealed (e.g., "Killer Dolphins Slaying, Sexually Assaulting Porpoises in San Francisco"[22]), bottlenose dolphins have recently garnered a reputation for attacking and killing harbor porpoises—a toothed cetacean species half their size. The official term for this behavior—coined in 2011—is *porpicide*.[23] Although the recent media blitz focusing on porpicide would give the impression that it's a novel occurrence, it's entirely possible that it has been happening for millennia without anyone noticing until now.

The first inkling scientists had that it was not all sweetness and light beneath the waves was in the early 1990s, when harbor porpoises from the Moray Firth on the east coast of Scotland were beginning to wash ashore with strange and gruesome injuries. Between 1991 and 1993, 63 percent of dead harbor porpoises found on the beaches displayed signs of being attacked by another toothed cetacean species.[24] These animals exhibited "multiple, internal, ante-mortem injuries characterized by extensive bruising and hemorrhage in the subcutis and underlying musculature, particularly over the head, dorsum, upper chest wall and flank."[25] The injuries were caused by blunt impact, sometimes so violent as to cause the porpoises' livers and lungs to rupture, and to fracture their spines. The occasional tooth marks left on the skin of these dead animals were examined to determine which species was responsible for these attacks. Only one species was a perfect match: the bottlenose dolphin. Tourists and researchers had witnessed aggressive encounters between bottlenose dolphins and porpoises in the Moray Firth before, but only now were the pieces of the puzzle falling into place to suggest how common—and deadly—these interactions might be. The researchers noted that "these interactions were highly-violent and non-consumptive."[26] In other words, the dolphins were not eating the porpoises—they were simply killing them. Violently.

Porpicide is not restricted to some rogue population of bottlenose dolphins in the Moray Firth however. Between 2007 and 2009, forty-four dead stranded harbor porpoises were found along the beaches between Humboldt County and San Luis Obispo County in California, with scientists confirming that the porpoises had "sustained traumatic injuries from interactions with bottlenose dolphins."[27] The injuries included a now familiar gruesome list of telltale bottlenose violence, including "multiple fractures of the ribs, spinal column, skull, scapula and tympanic bulla, and lung and soft tissue lacerations and contusions." Scientists in California also witnessed and documented three bottlenose dolphin attacks on harbor porpoises, one of which ended in the death of the porpoise. The attacks involved behavior such as:

> Ramming: Hitting the porpoise at fast speed with the rostrum and the side of the body, often repeatedly and sometimes by multiple animals at the same time.
>
> Tossing: Partially or completely throwing the porpoise out of the water with fast and violent maneuvers, using either the rostrum or the fluke to hit it. Often performed in sequence. This tactic produces loud noises as the porpoise is violently hit on both sides, and often sends the victim somersaulting out of the water.
>
> Drowning: Repeatedly lifting a dolphin's upper body out of the water at a 45° angle and letting it forcefully drop on top of the porpoise's head, pushing it underwater. Also positioning a dolphin's rostrum underneath the flukes of the porpoise and lifting the dolphin's head out of the water, effectively keeping the head of the porpoise underwater. Both techniques are effective in tiring and disorienting the porpoise, and in preventing it from breathing.[28]

The number of recorded porpoise deaths in California was so striking that the National Marine Fisheries Service declared the situation an *Un-*

usual Mortality Event, ultimately concluding that sustained and regular bottlenose dolphin attacks resulting in "blunt trauma" were the likeliest cause of the sudden drop in the harbor porpoise population.[29] Because porpicide appeared to occur with regularity in two separate populations of bottlenose dolphins over the course of many years, it couldn't be classified as aberrant behavior. Based on the number of dead porpoises washing ashore, and the estimated number of attacks that must be taking place that didn't result in the death of the porpoise, it's probably the case that attacking porpoises was in fact fairly common behavior for these dolphins. But what could be triggering the attacks? It was unlikely to be a territorial dispute; the porpoises were not competing for the same food source as the dolphins. Interestingly, the evidence suggested that the primary perpetrators of the attacks were all adult male dolphins; an important clue. Also, the porpoises being targeted in Scotland were not quite fully grown: between just 1.0 and 1.5 meters in length. So why would adult male bottlenose have an interest in killing medium-sized harbor porpoises? Scientists' best guess at the moment is that these dolphins are practicing their killing techniques, sharpening their skills for the day when their intended victims finally make an appearance: bottlenose dolphin calves. Bottlenose calves and medium-size porpoises are approximately the same size, so any skill acquired in killing porpoises can be easily transferred to killing calves when the time comes. An alternative hypothesis is that these male dolphins might be confusing the porpoises with bottlenose dolphin calves. Either way, the most likely explanation as to why porpicide exists is that it is a variation of a decidedly unfriendly and unpeaceful behavior: infanticide.

Infanticide—where adult dolphins kill newborn or young dolphins of the same species—has not been reported in the scientific literature through direct observation, although there is ample indirect evidence that it exists. Between 1992 and 1996, eighteen bottlenose dolphin calves (less than one year old) were found dead on the Moray Firth beaches (the same area where porpicide was reported), bearing the telltale signs of having been killed by larger dolphins. The scientists describe the gruesome state of these dolphin calves as involving

> multiple, internal, ante-mortem injuries including bruising around the head and thorax, multiple rib fractures with associated haemorrhage and bruising, ruptured lungs resulting from penetration by fractured ribs, and, in two cases, spinal dislocation. In each case the skin bore fresh parallel tooth marks but in no cases were parts of the dolphin eaten.[30]

As you likely noticed, this description of injuries is nearly identical to those found on the dead porpoises. Researchers in Scotland witnessed an adult male dolphin attack a newborn calf, separating it from its mother and nearly drowning it,[31] and also reported an observation of a male bottlenose dolphin attacking an already dead bottlenose dolphin calf—although they arrived too late on the scene to confirm if the male had actually killed the calf, or was just having a bit of fun after having found an interesting (grisly) plaything.[32] Dead bottlenose dolphin calves with severe internal injuries that were thought to be caused by violent attacks by other bottlenose dolphins have also been recorded in Virginia.[33] Nearly half of the necropsied calves that stranded dead on beaches in Virginia between 1996 and 1997 were thought to have been killed by adult dolphins. "It looks like someone had taken a baseball bat and just literally beaten these animals to death," the lead author, Dale G. Dunn, was quoted as saying.[34]

These reports constitute fairly convincing evidence that adult male dolphins had been involved in intraspecies infanticide. While infanticide most certainly does not fall under the umbrella of peaceful behavior, it is not at all uncommon in the animal kingdom. For dolphins, it has a very clear goal: kill the calf and the mother will be ready to mate in just a few days. Wherever you find a group of animals living under the right social conditions with males competing for access to females for reproductive purposes, you set the scene for infanticide. Lions are probably the most well-known example. If a male lion takes over leadership of the pride, he might kill the cubs sired by previous males, ensuring that the females come back into heat so he can father the next batch of cubs. As far as evolution is concerned, it would not be a good strategy for that male to spend years of his life providing for and defending cubs that don't carry any of his genetic material. Consequently, more than one quarter of all cub deaths can be directly attributed to infanticide committed by new males taking over a pride.[35]

The story for dolphins (specifically bottlenose dolphins) is slightly different. Unlike lions, male dolphins do not and cannot know if they are (or are not) the father of the calves in their social groups. This creates a sexual competition scenario where males attempt to mate with females whenever they get the opportunity (and possibly create opportunities by killing calves), and females devise strategies to make sure that they do not become impregnated by an inferior male, while also making sure that dangerous males are kept a safe distance from their vulnerable newborns. For some dolphin species, male mating strategies consists of finding ways to coerce (i.e., force) the females to mate with them.[36] For the male Indo-Pacific bottlenose dolphins in Shark Bay, Australia, males form short- and long-term alliances lasting anywhere from a few minutes to a few decades. These alliances often crop up when females are ready to mate; it is thought that the males team up in order to keep the females all to themselves—that is, prevent other males from getting the chance to mate with them, a behavior termed *herding*.[37] Sometimes two or more of the smaller alliance groups will team up to form larger groups in order to defend their females or steal females away from smaller groups of males. Male alliance groups often engage in open combat which can be extremely violent, with an observation of a male bottlenose dolphin that was knocked unconscious (and nearly drowned) while fighting with a coalition group of three other males.[38]

A similar social structure has been observed in the bottlenose dolphins of Sarasota Bay in Florida. These dolphins also live in similarly fluid social groups, with females banding together in nursery groups, and males forming long-term alliances (over twenty years in some cases)—usually in pairs.[39] The male alliances are, like Shark Bay, thought to be a strategy for gaining access to reproductive females. The female groups likely form to help mothers raise their calves; protecting them from danger (possibly in the form of sharks or infanticidal male dolphins), and making it easier to find/hunt for food.[40]

In order to exert some control over mating, female dolphins may ovulate at random times or mate with lots of males so as to make it impossible for the males to guess who the father might be.[41] It has been suggested that the extra flaps of tissue observed in the vaginal walls of some species of dolphin (sometimes called a pseudocervix) might allow the female to control which male's sperm fertilizes her egg. The constant struggle between male and female interests (all boiling down to sex and who gets to father the calf) is an important factor driving the structure of many dolphin societies—as it is for most social mammals. But I do not want to leave you with the impression that infanticide—or even mate access—is the driving mechanism behind the structure of all dolphin societies. That's not likely true. Ecological factors like the need to band together to hunt for food or avoid predators are just as important in explaining dolphin group structure as reproductive factors.[42] In some dolphin societies, competition between the sexes and the threat of infanticide might be minimal to non-existent. There is, for example, a well-studied group of around eighty bottlenose dolphins in Doubtful Sound, New Zealand that exhibit the typical complex social groupings seen in other dolphins, with numerous long-term relationships.[43] However, these dolphins, unlike other bottlenose dolphin populations, form mixed-sex groups (observed over 90 percent of the time), where mate competition is negligible. Males and females formed long-term partnerships with each other—something quite rare for bottlenose dolphins. Researchers studying this group suggest that the need for dolphins to share information about where to find food in the inhospitable fjord where they live might generate a higher degree of cooperation than that seen in other bottlenose groups. These same-sex groupings might also be explained by the females having the upper hand when it comes to choosing a potential mate,[44] without having to worry too much about being coerced or herded. Other bottlenose dolphin societies—like those in the Moray Firth in Scotland,[45] and the Shannon Estuary in Ireland[46]—retain the typical fluid social structure, but also lack the kind of male–male alliances or female bands seen in the Shark Bay and Sarasota groups. It should also be noted that even for the more antagonistic Sarasota and Shark Bay dolphins, mixed-sex groups are seen frequently enough: over 50 percent of the time in Shark Bay and 31 percent of the time in Sarasota. So it's not the case that males and females never get along, and that adult males always have infanticide on the brain. But, as the Moray Firth dolphins have shown, infanticide likely remains an attractive mating strategy for dolphins no matter what the particulars of their social structure.

Other aggressive acts

Female dolphins sometimes engage in disciplining of their calves,[47] which often incorporates fairly unpleasant—if not violent—behavior. A commonly cited form of calf discipline involves the mother pinning the helpless calf on the seafloor,[48] which, for an animal that needs to be at the surface to breathe, is likely to be either extremely uncomfortable or terrifying. Spotted dolphin mothers have been observed repeatedly knocking their young calves into the air,[49] and calves sometimes exhibit signs of having been bitten by their mothers.[50] Bottlenose dolphin mothers sometimes produce an aggressive vocalization termed a "thunk" to warn calves that stray too far from their side.[51] These kinds of harsh parenting techniques are fairly ubiquitous in the animal kingdom (which can be confirmed by watching how some human parents deal with misbehaving toddlers), but certainly challenge the idea that dolphins are unusually peaceful when it comes to handling a precocious youngster.

Aside from the porpicide issue, there is evidence that the idea of interspecies harmony is really a bit far-fetched. Dolphin species regularly interact in non-violent and even cooperative ways with other marine mammal species,[52] but for every example of a friendly interspecies interaction (e.g., humpback whales and bottlenose dolphins playing games together,[53] dolphins and dugongs socializing[54]), you can find an example of aggressive interspecies interaction (e.g., bottlenose dolphins attacking a Guiana dolphin calf,[55] or spotted dolphin and bottlenose dolphin groups fighting[56,57]). For example, striped dolphins, Risso's dolphins, and common dolphins in the Gulf of Corinth are almost always sighted together in mixed-species groups, where both affiliative and agonistic encounters have been observed. The scars on the bodies of these animals testify to the regularly aggressive nature of their interspecific interactions.[58] White-beak dolphins were likely to have caused the scarring found on the skin of two harbor porpoises found stranded on a beach in Belgium,[59] and Pacific white-sided dolphins have been observed harassing (and possibly trying to play with/drown/kill) a newborn harbor porpoise.[60] And while Commerson's dolphins and bottlenose dolphins have been observed feeding peacefully together in Patagonia, the peace can be interrupted by occasional violent interactions.[61] And those same Moray Firth bottlenose dolphins responsible for killing harbor porpoises have been implicated in the deaths of a juvenile pilot whale, a juvenile Risso's dolphin, and adult striped and common dolphins—all found stranded with internal injuries likely resulting from another series of bottlenose attacks.[62] These observations challenge the infanticide hypothesis as the only or primary explanation as to these animals' violent behavior insofar as these other species are much larger than harbor porpoises or bottlenose calves.

Dolphin-on-dolphin aggression has been witnessed in captive dolphins, which should also be mentioned. Granted, there is a perfectly valid argument to be made that captivity itself might elicit—or at least exacerbate—aggressive behavior in dolphins (due to unnatural social groupings, stress, boredom, etc.),[63] but these examples should at least be entered into the record. Animal husbandry reports penned by veterinarians confirm that dolphin aggression is a common problem, and recommend a number of strategies for coping with it;[64] usually separating animals that antagonize each other. Some dolphins just don't get along, and regular physical attacks and threats persist until something is done to change the structure of the social group.[65] One of the major problems appears to be that captive groups establish and maintain a strict social order,[66] with dominant individuals (both male and female) doling out punishment on submissive dolphins (either of the same or different species).[67] It's not always the case that the bigger dolphin is the dominant one—bottlenose dolphins have been known to maintain dominance over much larger killer whale tank mates.[68] Dominance and rank is not something that has been found/studied in wild dolphin populations,[69] and indeed might play no part in wild dolphin societies.[70]

Perhaps the most well-known—and highly publicized—aggressive encounter between captive dolphins is the death of the killer whale Kandu at SeaWorld San Diego in 1989. Kandu sustained a fatal injury while engaged in an aggressive encounter with another female killer whale, Corky. The incident was described as a "normal, socially induced act of aggression to assert her dominance" by SeaWorld officials.[71] Unfortunately, this act of aggression fractured Kandu's jaw, causing massive hemorrhaging, which killed her in a matter of minutes.

Not all aggressive encounters need be so dramatic. In fact most of the aggression that is witnessed in dolphins, both in the wild and captivity, does not even contain physical contact, but it's aggression nonetheless.[72] Just like we see in humans, aggressive displays involving postures and vocalizations can indicate an animal's internal emotional state. For a dolphin, aggressive postures include an open mouth (sometimes a play behavior, but very often a sign of stress or agitation), an S-shaped body posture, head bobbing or shaking, charging, and flaring pectoral fins out to the side (in order to appear larger and more threatening).[73,74,75,76] Vocalizations include loud pops, squeaks, buzzes, jaw claps, jaw pops, tail slaps, etc.[77]

Even when aggressive physical contact does occur, it need not result in the kinds of dramatic injuries witnessed in porpicide or Kandu's death, and can include ramming or hitting with the rostrum, flukes, or peduncle which otherwise leave no obvious physical trace of having occurred.[78] Perhaps the most obvious sign of non-fatal aggression in dolphins is the presence of tooth rake marks—caused when a dolphin bites or rakes its teeth across the flesh of another dolphin. This sometimes occurs in the context of play, but quite often is the result of both inter- and intraspecific aggressive encounters.[79] The presence and distribution of tooth rake marks on an individual dolphin or a population of dolphins is used by scientists to determine the prevalence of aggressive behavior.[80] Species like Risso's dolphin are often completely covered in rake mark scars from aggressive interactions with conspecifics, with the scarring itself being a signal of the male's fighting prowess, and thus his quality as a mate.[81] The same is true of the boto, or Amazon river dolphin. Were it not for a body completely covered in scars built up after a lifetime of aggressive interactions with conspecifics, adult male botos—which are almost entirely pink from scar tissue—would likely retain their original gray coloring.[82] In fact, botos have a reputation for some rather serious male–male squabbles, with scientist Tony Martin noting that "the males beat the hell out of each other. They are brutal. They can snap each other's jaws, tails, flippers, lacerate blowholes. The large males are literally covered with scar tissue."[83]

How common is aggressive behavior in dolphins?

Up to now we've been dealing with a collection of anecdotes and observations that suggest that dolphins are not always peaceful animals. But it would be handy to have an objective means of calculating just how often dolphins exhibit friendly or unfriendly behavior to determine how "abnormal" aggression might be. Scientists studying animal behavior sometimes create a list of all of the different behaviors exhibited by an animal, called an *ethogram*, and also a *behavioral budget*, which calculates the frequency at which different behaviors are observed. By looking at ethograms and behavioral budgets, we can get an idea of just how common friendly or aggressive behavior is.

For one study of behavioral development in bottlenose dolphin calves in Shark Bay,[84] nine newborn dolphin calves were observed for the first ten weeks of their lives as closely as possible. These dolphins were involved in provisioning—that is, that were occasionally fed by humans. The average rate of aggression when the mothers were in the provisioning area was 0.25 aggressive behaviors per hour, but just 0.01 aggressive behaviors per hour when not being provisioned. As the authors of this study pointed out, this provides some evidence that provisioning the dolphins (i.e., human contact/influence) does seem to boost aggression rates. A study of two captive mother–calf pairs observed for a total of thirty-two hours produced an ethogram where about four of the fifty-one behavioral categories were aggressive in nature.[85] These two studies suggest that aggressive behavior witnessed in young calves (and their mothers) exists, but is not all that prevalent.

When looking at ethograms of wild spotted dolphins in the Bahamas for the first four years of their lives, a bit more aggression crops up.[86] Of the forty-six behavioral events included in this ethogram, about eight could be considered aggressive. These included relatively harmless displays like jaw claps and tail slaps, but also things like forcibly shoving the rostrum into the genital slit. Aggressive encounters involving disciplining (e.g., pinning the calf to the seafloor) were rather rare—observed just twenty-three times in the course of the study. Nonetheless, a large percentage (about 17 percent) of behaviors listed in this ethogram were aggressive in nature. Only about 11 percent of the behaviors could be placed in the "friendly/gentle/peaceful" category, with the rest belonging to non-social activities like foraging, feeding, resting, swimming, etc. So for this ethogram, even though it does not provide us details with frequency of observed behaviors, we find a larger variety of aggressive than friendly behavioral categories.

Looking at an ethogram compiled from observing a wild Atlantic spotted dolphin population that included all possible age classes, aggressive behavioral categories take up a similar percentage of the total as the previous study.[87] Of the 107 behavioral categories listed, about nineteen (i.e., 18 percent) could be considered overtly aggressive, and about eleven (i.e., 10 percent) overtly friendly. Even though these results do not tell us how frequently the animals engaged in aggressive or friendly behaviors, we can at least state that aggressive behaviors constitute a sizeable chunk of the behavioral repertoire displayed by spotted dolphins. In a catalog of vocalizations produced by spotted and bottlenose dolphins in the Bahamas, one study labeled over 50 percent of vocalizations as occurring during agonistic and aggressive behavioral exchanges, either between dolphins of the same or different species,[88] suggesting that dolphins spend a fair amount of time making hostile sounds directed at each other. And using tooth rake marks as an indicator of the prevalence of aggression, it was found that 83 percent of bottlenose dolphins in Shark Bay bore the telltales signs of having been involved in an aggressive encounter, with the males in the population far more likely to have rake marks. Although intraspecific aggressive interactions among killer whales might be rare for the most part, with only the occasional report of intergroup fighting,[89] studies of tooth rake marks on a population of killer whales in New Zealand reveal some extensive scarring from aggressive interactions on a couple of individuals.[90]

It should be noted that scientists are forced to make a somewhat subjective decision as to if observed behavior falls into the category of aggressive or not. Although it might seem pretty straightforward to label a scenario where a group of male killer whales are body slamming and biting each other as an "agonistic encounter," it's almost impossible to know where to draw the line between vigorous play behavior meant to hone fighting skills and true aggression. Even simple behaviors like *tail slap* are not necessarily easy to categorize; is this always an aggressive act? Would it be better to categorize it as an act of frustration, or a warning signal? Is it also used during play? Context always changes the meaning of a signal, so it's not necessarily correct to just lump this under the category *aggressive* every time you observe it. Objective measures like increased levels of stress hormone, traumatic injury, or death are very likely to be good indicators that a dolphin is experiencing something unpleasant, but these things are difficult to measure for wild dolphins.

Even if we allow for a wide margin of error in terms of the definition of *aggressive behavior* because of the shortcomings I just mentioned, these ethograms and behavioral budgets certainly demonstrate that aggression is a common component of the complex repertoire of dolphin behavior. And when we tack on the issues of infanticide and porpicide, it is hard to arrive at the conclusion that dolphins always live in "harmony in their environments," and "enjoy harmony with other species." If anything, bottlenose dolphins could even be considered unusually aggressive animals, with dolphin experts Richard Connor, Randall Wells, Janet Mann, and Andrew Read noting that "bottlenose dolphins are also one of the very few mammals known to direct lethal, nonpredatory aggression at other mammalian species."[91] Finally, it goes without saying that the many species of aquatic animals that dolphins hunt, kill, and consume as food (e.g., mackerel, seals, gray whale calves) would, if allowed to chime in on the question of harmony, forcefully argue that dolphins can be fairly nasty individuals. There are, after all, no vegetarian dolphins.

Complex Social Behavior

After reviewing the above evidence, it certainly does not seem that dolphins should be classified as peaceful animals that live in harmony with other animal species. But this does not necessarily negate, and perhaps only enhances, the idea that dolphins are unusually complex animals when it comes to social behavior. From stories of dolphins mourning their dead or banding together to defend each other from sharks, dolphins' social behavior is often considered particularly human-like in both its complexity and its occasionally empathetic nature. But does the scientific evidence support the idea that dolphins lead unusually complex social lives?

Complex social structures

Aquatic mammal species live in an environment dissimilar to their terrestrial cousins in one very important aspect: in the ocean, there is nowhere to hide. Danger in the form of undetected predators (e.g., sharks, killer whales) lurking in the deep is something (smaller) dolphins have had to deal with twenty-four hours a day for the past few million years. Aside from the occasional chance to seek refuge in relatively safer shallow waters, dolphins' primary solution to this problem, like many species of fish, is group living. But unlike fish, dolphin social groups are influenced by a number of other ecological pressures, including the need to forage cooperatively (for some species), and the need to navigate complex social relationships focused on competition between the sexes. The result of the tension between the need to work together for protection and access to food, and the need to compete with each other for access to mating partners has resulted in dolphin societies being some of the most varied and complex in the animal kingdom.[92] This is not to say that other mammals like meerkats, elephants, wolves, and chimpanzees, or avian species like crows or parrots do not live in complex social groups. They most certainly do. But dolphins exhibit particularly complex—and even unique—social systems.

The societies of many dolphin species throughout the world (e.g., bottlenose dolphin, Guiana dolphin)[93] are often described as consisting of fission–fusion groupings. This means that individual dolphins spend time in any number of different social groups throughout the day, with group composition constantly changing.[94] This is a similar scenario seen in some primate and elephant species,[95,96] and is often compared to chimpanzee social structure.[97] But in some parts of the world, the complexity of these interweaving and ever-changing relationships for dolphins has reached impressive levels of complexity. We have already seen that male dolphins in Shark Bay and Sarasota form long-term alliances as a means of gaining access to females. "First-order" alliances involve two or three males that spend almost all of their time together, with partnerships that can last their entire lives and considered some of the strongest social bonds observed in the animal kingdom.[98] Occasionally, two first-order alliance male groups will join together to form a second-order alliance. These second-order group combinations can last up to fifteen years, and are a handy means of defending or taking females from other alliances.[99] On occasion, multiple alliances will form a super-alliance[100] or a third-order alliance[101] consisting of second-order alliances teamed with other first- or second-order alliances. These larger alliances involve constant switching of smaller group alliance memberships along non-random lines while working together to herd females or engage in fighting, suggesting that they maintain their own rules about which members are allowed to join forces, and indicating a complex internal social structure involving "nested alliances" within a larger social network.[102]

Male and female dolphins in Shark Bay are unlike many social mammal species in that they do not defend territorial ranges, and young dolphins will disperse to form social groups that are not based on kinship. In contrast to all other species expect humans, the hundreds of dolphins in Shark Bay do not live in closed social groups (typically based on kinship for other species) that either ignore or tolerate each other (like gorillas) or actively fight each other (like chimpanzees) when they come in contact. Instead, they live in an "open society," with all animals in the area interacting, socializing, and forming alliances with individuals from overlapping ranges.[103] There are many other examples of complexity within social structure of dolphin species, like the matrilineal groups of killer whales with their vocal dialects resulting from social learning,[104] but it's the fantastically complicated social networks observed in Shark Bay that stand out as being unusual in the animal kingdom, and provide a strong argument for the idea that dolphins lead extraordinarily complex social lives.

One final note concerning the supposed uniqueness and/or complexity of dolphin social structure as it pertains to socio-sexual behavior. It is not the case that dolphins are the only species other than humans to have sex for pleasure,[105] or engage in same-sex sexual behavior[106] or coercive mating strategies (often incorrectly described as "rape"), nor is there any evidence to support the surprisingly popular notion that dolphins insert their penises into each other's blowholes as a means of sexual stimulation (or for any other reason).[107] These popular culture memes stem from a misinterpretation of the scientific literature, which suggests that dolphin socio-sexual behavior is no more complex—or bizarre—than that of other social mammals.

Cooperation and altruism

One of the factors driving group living in dolphins is the benefit it creates for both individuals within the group, and the group itself. Mutualistic interactions can, for dolphins, take the form of cooperation (e.g., cooperative foraging), but also a paradoxical kind of behavior called altruism. Altruism involves one animal helping another animal, but at a potential cost to itself (i.e., at a cost to its reproductive fitness), possibly even at a risk to its own life. The evolution of mutualism, and particularly altruism, has been a hot topic of debate in evolutionary biology ever since Darwin grappled with the problem in *On the Origin of Species*. Many hypotheses are on offer as to what exactly is driving the evolution of prosocial or eusocial behavior in the animal kingdom. As recently as 2012, a squabble erupted between supporters of the inclusive fitness via kin selection hypothesis (e.g., Richard Dawkins)[108] versus the multilevel selection and/or group selection hypothesis (e.g., E. O. Wilson).[109] Regardless of which hypothesis best explains its evolutionary origins, examples of cooperative and altruistic behavior for dolphins abound.

Cooperative behavior

The Shark Bay and Sarasota bottlenose dolphin alliances are an obvious example of cooperative behavior, whereby groups of male dolphins work together to herd females and thus increase their chances of siring a calf.[110] There is also strong evidence that many delphinid species forage in a highly coordinated and cooperative manner. Dusky dolphins in Argentina are famous for their ability to coordinate their foraging behavior so as to force large schools of fish to form giant "bait-balls," making it easier for individual dolphins to dart in and capture fish.[111,112] Spinner dolphins employ a similar means of cooperating to herd prey, resulting in orderly turn taking and precise timing during feeding events.[113] In Cedar Key, Florida, bottlenose dolphins hunt using a technique in which a "driver" dolphin will herd fish using fluke-slaps toward a group of non-driving dolphins lined up to form a barrier preventing the fish from escaping.[114] The fish then leap out of the water, and the dolphins grab them out of the air. Dolphins in Bull Creek, South Carolina perform a coordinated foraging behavior called "strand-feeding."[115] By charging toward shore in unison after schools of fish, the fish end up trapped on the muddy beach. The dolphins wash up alongside the fish, and gobble them up as they slowly slide back into the water. Even humans and dolphins have been known to coordinate together to capture fish, with dolphins off the coasts of South America,[116] Australia,[117] India,[118] and Africa[119] helping to drive fish into the nets set by their human assistants, allowing them to capture any fish that might escape. Dolphins have even been observed sharing food, with adult false killer whales and rough-toothed dolphins offering captured fish to younger dolphins.[120]

Transient killer whales (that feed on marine mammals) also hunt cooperatively, with studies showing that hunting group size usually consists of between three and four animals; the ideal number to maximize/balance foraging efficiency versus energy intake from their kill.[121] The coordinated hunting efforts of one mammal-eating killer whale type called "pack ice killer whales" have become a hit on the internet, with videos showing the complex coordinated behavior called "wave-washing."[122,123,124] Once these killer whales spot a seal or other prey species taking refuge on an ice floe, they will swim some distance away and then turn in formation and swim rapidly toward the ice floe. This results in a wave forming that washes the seal off the ice and into the water where the killer whales can then pursue it. Many species of dolphins fan out to form lines or ranks when feeding, which some consider a type of cooperative feeding,[125] although others have pointed out that this tactic might not be a sign of cooperation, but a means of preventing/mitigating feeding competition.[126] Nonetheless, ample evidence exists that dolphins can and do cooperate to both find and capture prey.

Altruistic behavior

Dolphins have long been celebrated for their habit of aiding drowning human swimmers. Providing assistance or care to another animal like this (of either the same or a different species) is a form of altruism called *epimeletic* behavior, and falls into two categories: *nurturant* (i.e., providing care to young animals), and *succorant* (i.e., providing care to animals in distress).[127] In a review of the literature on altruism and epimeletic behavior in dolphins from 1982, Richard Connor and Ken Norris stated that "the evidence of epimeletic behavior, though based wholly on anecdotes, is so common as to be overwhelming in its broad detail."[128] In the thirty years since this review was published, ample well-documented evidence of epimeletic behavior in dolphins has been added to the scientific literature, confirming that it is likely to be both common and widespread for delphinind species.

Adult dolphins (especially mothers) take an active role in caring for their young, resulting in widespread nurturant behavior being reported in the scientific literature. Like other mammal species living in complex societies where young animals acquire social (and possibly cultural), cognitive, and motor skills (e.g., elephants,[129] chimpanzees[130]), bottlenose dolphins go through a protracted period of development, with calves sometimes continuing to suckle for three to five years and, in some cases, staying close to mom for up to a decade.[131,132,133] In the case of some killer whales, calves will sometimes remain in their mother's family group for life—rarely separating from their matrilineal group for more than a few hours.[134] Mothers provide protection for their calves, assist in foraging, and are generally in close physical contact with their calves—especially for the early stages of their lives. Female killer whales go through menopause in their thirties or forties, but remain active in their social groups into their eighties and nineties. It's extremely rare in the animal kingdom for females to go through menopause (only humans, killer whales, pilot whales, and possibly sperm whales do it),[135] but a long-term study of killer whales from British Columbia has shown that these killer whale grandmothers continue to fulfill a vital role in the group even without the ability to reproduce.[136] Male (although not female) killer whales' chances of dying are increased anywhere from threefold (if they are under thirty years old) to eightfold (if they are over thirty years old) in the event of their mother's death, suggesting that these matriarchs have a rather large impact on their sons' lives. It's not known how these mothers and grandmothers are aiding their sons (e.g., foraging help, protection during agonistic encounters), but the relationship between the presence of older females in a group and the survivability of adult male killer whales is rather astonishing. These females are doing more than just raising their calves until maturity; they continue to aid their offspring through their entire life cycle.

Alloparenting, where individuals other than the calf's mother help in parenting tasks/care,[137] is also widespread in dolphins, with instances of females "babysitting" calves for extended periods of time or even "adopting" orphaned dolphins, including calves being cared for/adopted by dolphins of a different species.[138,139,140,141,142] Unlike other social mammalian species (including some primates),[143] adult male dolphins do not typically engage in very many nurturant behaviors directed at young dolphins,[144] as you might have guessed given the previous discussion on infanticide.

A handful of the many examples of succorant behavior include a bottlenose dolphin mother helping her calf to swim after it had been paralyzed by a boat propeller strike,[145] and sick animals being supported at the surface by their companions for extended periods of time.[146] Evidence exists that dolphins sometimes (but not always)[147] band together to fend off shark attacks.[148,149,150,151] Dolphins of both the same and different species have been observed coming to the aid of individuals being captured/killed by humans—even biting restraining/harpoon lines attached to their struggling compatriots.[152] Dead calves being carried/supported by their mothers or other dolphins for extended periods of time (e.g., more than two days) has been widely witnessed and reported.[153,154,155,156,157,158] Dolphins have been observed remaining close to dead social partners/family members for days on end,[159] and even chasing away other dolphins or curious humans who get too close.[160,161] One of the explanations as to why dolphins beach themselves in groups is that the group will follow sick/injured animals into shallow water (a safer place to convalesce than the open ocean), and refuse to abandon their vulnerable companion/family member when the tides change, resulting in the entire group being stranded on the beach.[162] Healthy individuals refloated after a stranding are often observed restranding themselves so as not to be parted from their sick/injured compatriots still on the beach.

The cooperative, epimeletic, and altruistic behavior that has been observed in dolphins is by no means unique to delphinind species. In fact, these kinds of behavior are fairly ubiquitous in the animal kingdom, which is why they've been at the heart of the debate between kin versus group selection for so many decades. Coordinated group hunting—including interspecies coordination—has been observed in hundreds of species, from mammals to fish, insects, and spiders.[163] Attending to and supporting animals that are injured or have died has been observed in many species, including sea otters, manatees, sea lions, harbor seals, baboons, langurs, orangutans, chimpanzees, gorillas, elephants, and ungulates.[164] And while banding together to fend off sharks appears at first glance to be a unique and complex form of altruistic behavior, many species, ranging from birds to rodents to fish to primates, engage in mobbing behavior whereby they cooperatively attack predators in their environment in small or large groups. Gulls,[165] meerkats,[166] barn swallows,[167] robins,[168] marmosets,[169] ground squirrels,[170] damselfish,[171] and bluegills,[172] are just a handful of species that, like dolphins, join forces to repel predators. Social insects like ants, bees, or wasps are perhaps the most altruistic of all animals, with (for some species) all members of the colony dedicating and often sacrificing their lives in an effort to care for and protect their queen, the young, and the colony as a whole. Only a handful of mammals (e.g., the naked mole rat) show a similar level of eusocial or prosocial behavior, which has been dubbed "extreme altruism."[173]

A few noteworthy examples of other species engaging in cooperative and altruistic behavior similar to what is observed in dolphins include groupers and moray eels, which join forces to hunt together, making it easier to overwhelm and capture their intended prey.[174] Aggressive interactions between these species while hunting cooperatively has never been witnessed. Adult common swifts will come to the aid of young fledglings embarking on their first flight attempts.[175] The adults will, sometimes in groups, fly under the young swifts, helping to keep them airborne by propping them up. Swifts have even been observed providing this kind of flight assistance to other species of birds (e.g., the house martin). Young iguanas often stay together in close-kin groups for the first few months of their lives. Experiments have shown that in a simulated predation attempt whereby a "hawk" was flown over the group, male iguanas will often spring into action, throwing themselves on top of the females at the moment the hawk appears to attack.[176] This altruistic act might well protect the females, but most certainly endangers the lives of the selfless males.

When humans witness these kinds of prosocial behavior in dolphins and other animals, it immediately brings up the question of what is going through their minds. Are mother dolphins showing human-like empathy or grief when they support their calves' lifeless bodies? Many would argue that this is indeed the case,[177] although as we've seen, it is difficult if not impossible to ascertain the truth behind this sentiment scientifically. Other examples of epimeletic behavior in dolphins help to confuse the issue: dolphins have been observed carrying the bodies of dead sharks[178] and sea lions.[179] Why would they do this? Some scientists suggest that supporting a dead calf (or other animal) is a "misuse of an adaptive response" with the mothers simply unable to abandon the body or ascertain that the calf is dead.[180] Others suggest that it might be confused and neurotic behavior caused by being consumed with grief, something witnessed in humans as well.[181] But not all responses to injured or dead animals elicit human-like grief responses in dolphins. Sometimes, mothers seem to have no interest in attending to their dead calves.[182] And why, when dolphins were seen gathered around the body of their dead comrade in what appeared to resemble a vigil or funeral-like behavior (similar to what has been observed in scrub jays),[183] did researchers also witness males with erections sometimes directed at the dead animal?[184] This behavior is difficult to parse in human terms, which is why comparisons between human and dolphin epimeletic and altruistic behavior can have limited heuristic value.

We do not need to suggest that dolphins display human-like altruistic or cooperative behavior to nonetheless state that altruistic and cooperative behavior is common for dolphin species. But are these behaviors unique to dolphins, or in some way special? This is difficult to judge, but my personal opinion is that they are not. Given the ubiquity of these types of behaviors for such a variety of species (from mammals to lizards to insects), it's perhaps wrong to describe acts of altruism or cooperation by dolphins as anything other than a relatively normal suite of behaviors that can be observed in a great many animals.

Do dolphins lead unusually complex and peaceful social lives or not?

It is certainly the case that dolphins can be (at times) friendly and peaceful animals. And dolphins do indeed lead complex social lives. Included under this heading is a dizzying array of aggressive behavior including infanticide and interspecies violence. Dolphins also engage in cooperative and altruistic behaviors. Taken together, this evidence suggests that, while dolphins are deeply social animals, the idea that dolphins live in unusual peaceful harmony with each other and their environment is simply not correct. Neither is it correct that dolphins are unusual with regards to their epimeletic or prosocial behavior insofar as many animal species engage in similar behaviors. In reality, dolphins display exactly the right mix of aggressive, friendly, and complex social behavior that one would expect of any long-lived mammal that lives in large, complicated social groups. In this sense, dolphins are very similar to human beings. There is no doubt that humans can be amazingly generous, peaceful, loving, kind, etc. But there is also no doubt we can be cruel, hateful, malicious, and downright evil. It would be misguided to label the entire human species as either intrinsically peaceful or intrinsically cruel. Most personal philosophies and religions seem to agree that humans, like dolphins, are a little of both.

Why is it that Lilly's idea of the peaceful dolphin is so pervasive in modern culture? It's difficult to say why so many of us turn a blind eye to the sometimes violent behavior of dolphins; why we whitewash their behavior and continue to label them as peaceful, despite the overwhelming (and freely available) evidence that they can be pretty vicious. As much as I think that some New Age authors sometimes come across as naïve in their writings about dolphin behavior by glossing over the unpleasant stuff or dismissing reports of dolphin aggression as anomalous, I seriously doubt they are being disingenuous about their belief that dolphins are supernaturally friendly just to sell books or eco-tours. To be fair, dolphins can be extremely curious and friendly toward human beings, so there is plenty of material for them to work with. But this is not the whole story, and it does not do justice to the true complexity of dolphin behavior by failing to mention the less than peaceful side of them. Some modern authors, notably those that rely more on empirical evidence than personal experience (like Thomas White),[185] are keen to incorporate the well-documented aggression observed in dolphins into their description of the dolphin disposition, providing a more balanced, and consequently more accurate account. To truly understand dolphins, one must be open to the idea that they can be jerks.

I very much agree with Toni Frohoff's statement that "we're blinded to [dolphins'] emotional depth by our preconceived ideas of what they are—and what we want them to be."[186] If you truly believe that dolphins live in peaceful harmony with each other and are not capable of aggression,[187] then you will interpret dolphin behavior in such a way that it fits your version of reality, no matter what sort of dolphin-initiated atrocity you are witnessing. I will close this chapter with my own anecdote that emphasizes this point.

I was conducting research on wild Indo-Pacific bottlenose dolphins[188] from an eco-tour vessel when we encountered a male dolphin displaying classic signs of aggression for a dolphin: vigorously shaking his head, snapping his jaws together, blowing big bubble clouds, and charging and ramming other dolphins. At the time I was snorkeling with a group of eco-tour passengers. As I was nervously filming this behavior from a reasonably safe distance, one of the tourists swam up to the agitated dolphin to get a closer look. I tried to warn her of the danger she was in—there was every reason to believe that this aggressive large male dolphin would just as easily charge a human as another dolphin. I waved my arms and screamed in my snorkel, but the tourist swam straight into the fray. Luckily, the dolphins departed before anyone got hurt.

After we were back on the boat, the snorkeler told the others of her exciting encounter with the "friendly and playful dolphin." The other researchers, the boat captain, and I tried to explain to everyone that this was in fact a fairly dangerous encounter, and that the dolphin was acting quite aggressively. It was clear from the tourists' reaction that they did not believe us. As far as they were concerned, they had just witnessed the playful antics of the ocean's friendliest animal. Danger? What danger! Couldn't I see the dolphin's playful smile?

> Oh, you mean the smile that is a result of the shape of the dolphin's rostrum and cranial bones and is permanently frozen in place because dolphins can't change their facial expressions? The smile that the dolphin had as it rammed that other dolphin and probably broke its ribs?

Yes indeed, that smile.

7

The Deconstructed Dolphin

I hope we'll start to think more about what animals can do and less about what they can't.[1] *Temple Grandin and Catherine Johnson*

The Mammal Behind the Myth

With a generous helping of the scientific findings concerning dolphin behavior and cognition now heaped on the table, it's time to provide a quick summary of where the dolphin intelligence myth stands in relation to the evidence. There is solid scientific evidence to conclude that dolphins perform well on a number of cognitive tests associated with human-like intelligence. They have large and structurally complex brains, and an ability to comprehend some human-generated symbols and symbol-systems in experimental contexts. Dolphins live in unusually complex social structures, have emotions, display some form of self-awareness, and engage in a form of social learning that fits some definitions of culture. They engage in complex play, altruistic, epimeletic, and cooperative behavior. They've been observed using tools, and show some ability to plan and solve problems. But not all of the scientific evidence provides us with a clear idea of what's going on inside dolphins' minds. The evidence is too ambiguous, equivocal, or controversial to allow us to conclude that dolphin brains are so large or structurally complex that they support the same kind of cognitive complexity we see in human beings. Nor can we be sure that the presence of von Economo neurons or other brain structures are evidence of advanced emotional and social capacities in dolphins. And we don't know if self-awareness is similar to that of human self-awareness, if dolphins experience their emotions in a similar way to human beings, or if they possess complex emotions like empathy. Given the ambiguous state of the science at the moment, we can't say for sure if their memory, ability to plan, use tools, or solve problems, or their capacity for social learning and culture is complex and/or rare in the animal kingdom. And at this moment in time, the scientific evidence is either lacking or openly contradictory to conclude that dolphin communication is sufficiently complex to be considered equivalent to human language, or that they possess an advanced language-like communication system that we are going to one day decode. There's no evidence to suggest that dolphins communicate by transmitting three-dimensional holograms via their echolocation or have psychic abilities. There is also no evidence showing that dolphins have an awareness of the minds of other animals that is equivalent in complexity to the human capacity for theory of mind. It's also not correct that dolphins are unusually peaceful animals that live in harmony with other animals in their environment, and that they display altruistic, epimeletic, and cooperative behavior that is rare in the animal kingdom.

This then is the final result of the dolphin deconstruction process, and I hope I've attained my goal of providing an accurate and impartial overview of the current state of the science. I consider it something akin to airing science's dirty laundry; an overview of just how little we really do know when it comes to the science of the dolphin mind and dolphin behavior as it pertains to a discussion of intelligence.

Looking at dolphins' cognitive and behavioral abilities in the areas of symbol use, concept formation, sociality, self-awareness, problem solving, play, tool use, and culture, it is undeniable that they possess a suite of abilities that are not often found together in the animal kingdom. As many have pointed out, these skills are only seen in this combination in a handful of other species, including primates and corvids. Given that primates, corvids, and dolphins are distantly related on an evolutionary scale, the fact that they've converged on a similar set of cognitive skills provides us with a wonderful data set we can use to test our ideas of why these skills evolved—independently—in these lineages. Similarities in social structure and the cognitive skills required for complex group living are the best possible explanation at present, but this is by no means a settled issue.

But it's overly simplistic to state that primates, dolphins, and corvids occupy a unique spot on the scale of intelligence which considers them to be more human-like than other species. Consider the following four problems:

The extrapolation problem

Almost all of the evidence from the lab concerning the cognitive abilities of dolphins comes from the study of just one species: bottlenose dolphins (*Tursiops truncatus*). And the vast majority of the experiments on dolphin cognition were conducted on a regular line-up of dolphins (e.g., Ake, Phoenix, Toby). This includes studies of symbol comprehension, concept formation, self-awareness, etc. It is dangerous science to suggest that this species' or even these individual dolphins' skill in these areas can be extrapolated to either closely or distantly related cetacean species. Suggesting that minke whales, sperm whales, or common dolphins are likely to have similar levels of self-awareness because of results from the lab on mirror self-recognition involving just two bottlenose dolphins is equivalent to stating that spider monkeys can design Mars rovers because Stephen Hawking and Sheldon Cooper are good at math. It's entirely possible that MSR is limited to bottlenose dolphins (or maybe just two bottlenose dolphins), who might turn out to be the intellectual prodigies of the cetacean world.[2] Or it might be that species like the pantropical spotted dolphin are even more skilled, but we've just not subjected them to enough study in the lab. Either way, when we speak of dolphin (or cetacean) cognition at present, we're almost always referring to evidence derived from just a few representatives of a single species.

The overconfidence problem

As the two chapters on the study of the dolphin mind and dolphin behavior make clear, we are often left with more questions than answers where the results of studies on dolphin cognition are concerned. For example, why are dolphins skilled at symbol comprehension but not symbol production? Why do they pass the MSR test and display some forms of metacognition, but fail tests for invisible displacement? Why does dolphin tool use not involve tool production when they are known to be so good at manipulating things like bubble rings? When put in the wider context of the study of animal cognition in general, there is by no means a coherent picture as to how different cognitive skills appear to interact in the animal mind. Why for corvids, for example, are magpies capable of MSR, but not ravens, who are otherwise skilled at joint attention? Why is tool use more complex in rooks than ravens or magpies? And why are scrub jays so skilled at mental time travel and mental state representation but not tool use[3] or MSR? All of this suggests that scientists are still grappling with defining and testing for these cognitive skills, and unable to piece them together into a coherent framework resulting in linear scale, or a single intellectual capacity that governs performance on a variety of tests. For these reasons, it is foolish to make simplistic statements concerning what the results of specific tests of dolphin cognition mean about a dolphin's global cognitive world, or how closely their intelligence resembles that of a human. We simply have no definite idea as to what the nature of a dolphin consciousness, self-awareness, awareness of other minds, or emotional world looks like, nor how it might be driving their performance on specific cognitive tests. Therefore, we should be wary of overconfident statements to the effect that dolphins are displaying either a superior or unique kind of intelligence insofar as our understanding of dolphin cognition and how it compares to other species is rudimentary at best.

As the chapter on the dolphin brain highlighted, the evidence for dolphin intelligence based on properties and structures of the brain is so problematic as to be almost inadmissible evidence. There are lots of testable hypotheses concerning the size and structure of the dolphin brain and why it might have evolved to so closely resemble that of primates (in some respects), and whether or not this is the reason dolphin behavior sometimes resembles primate behavior. But these links are too fragile at the moment to allow for confident statement making on this particular subject.

The new evidence problem

It's almost inevitable that evidence coming from the study of other animal species will continue to reveal complex cognitive abilities on par with those of apes, corvids, and dolphins. Much of the evidence for dolphin intelligence exists because generations of scientists have been keen to find examples of it, having been inspired by Lilly's ideas of dolphin super-intelligence. Irene Pepperberg's research involving Alex the parrot reveals what can happen when a researcher is motivated and dedicated enough to design experiments that are fine-tuned to the behavioral, social, and perceptual needs of the species in question in order to elicit behavior that reveals their inner mental lives. Her work single handedly propelled parrots to the top of the list of intelligent animal species as far as symbol use was concerned, and similar levels of dedication and time invested surely have helped dolphins to achieve their many accolades. With the same level of innovation in experimental design applied to other species, keeping in mind the differences in perceptual and social worlds unique to each species, it is possible scientists will continue to find similar abilities in diverse species across the taxa. Scrub jays, for example, seem extremely skilled at episodic memory, mental state attribution, etc. which could very well result not from something unique to jay cognition, but because the researchers working with jays have been particularly clever in designing experiments meant to uncover these abilities. And bears have long been a neglected animal as far as cognition is concerned, and only in the last few years have researchers begun to reveal what is likely to be a level of cognition in bears that might yet rival that of corvids, dolphins, and primates.[4] The research involving insects, fish, and octopuses has also provided us with a laundry list of abilities that are unexpectedly primate-like, including tool use, teaching, social learning, numerosity, etc. It could well be that the many skills dolphins possess that seem extraordinary to us at the moment will eventually be understood as fairly ubiquitous in the animal kingdom in the near future.

The comparison problem

The elephant in the room is the fact that humans have cognitive abilities that allow us to do things that are unprecedented in the animal kingdom—to analyze and understand physical properties of the universe and create new materials from them, to create technology, cure disease, and send people into space. There is no question that no other animal species is able to accomplish these feats, which requires a specific kind of cognition that allows humans to manipulate and reason about information in complex ways. Thus, unless we discover that dolphins are building launch pads under the waves ready to send dolphin-astronauts into near-earth orbit, we will probably never reach a stage when we should consider dolphin intelligence as rivaling the intellectual abilities of an adult human. On the flip side, we're also never going to find a human that has a better working memory for numerical recollection than chimpanzees,[5] ability to navigate magnetic fields like a pigeon,[6] or echolocate with the same fine-scale resolution as a dolphin. If chimpanzees, pigeons, or dolphins were in a position to hold these "deficiencies" against us using their own cognitive capacities as the benchmark for intelligence, we would surely point out the inherent unfairness of the comparison being made. Human cognition, although based on a set of cognitive aptitudes that is shared by other species to some extent, is, by definition, unique to humans. The same is true of dolphin cognition. This highlights the problem of comparative intelligence versus comparative cognition, and suggests that the idea that the scientific evidence allows us to place dolphins on a linear scale of intelligence somewhere between the great apes and humans is, quite simply, wrong.

The Ethical Implications of a Deconstructed Dolphin

One of the aims of this book was to determine if the scientific evidence of dolphin intelligence was strong enough to form the basis for both legal and philosophical arguments for personhood in dolphins. It should be noted that many philosophical approaches to this problem do not necessarily reference the scientific evidence at all. Many animal rights advocates take it for granted that dolphins and almost all other animals are clearly "conscious enough" to warrant ethical treatment of some kind, and don't necessarily delve into (or perhaps even care about) the scientific debate as to how to define or test for animal self-awareness or animal consciousness. For philosophical positions advocated by Jeremy Bentham,[7] Peter Singer,[8] and Gary Francione,[9] the fact that an animal is sentient (i.e., can feel pain or pleasure regardless of the extent to which it might be self-aware or conscious/self-conscious) is enough to necessitate moral consideration and/or moral standing. Other philosophical approaches do lean rather heavily on scientific findings when building an argument for moral standing. Thomas White's position is that the current scientific evidence is strong enough to warrant personhood for dolphins, regardless of the many debates that cognitive scientists themselves might be having about the exact nature of (and how to test for) self-awareness, consciousness, emotion, etc. in animals.[10]

Scientist advocates like Marc Bekoff and Jonathan Balcombe suggest that even if researchers insist that there is no way to know for sure the full extent to which a dolphin is self-aware, conscious, or experiencing their emotions,[11] a responsible ethical stance should be to give them the benefit of the doubt and assume they are likely to possess these traits. This approach could minimize controversy insofar as it is based on the ambiguous state of the science, and should (if applied properly) avoid making a strong scientific claim about what goes on in animal minds in terms of subjective experience. Of course, this same argument can be applied to a great number of species that might be displaying behavior suggestive of self-awareness, consciousness, emotion, etc.

Not all arguments for dolphin rights hinge on how to interpret the ambiguity in the scientific evidence; many legal and philosophical approaches argue that the science itself *clearly indicates* that dolphins are special beings.[12] For example, *Wired Science* reported the following:

> Over the last several decades, researchers have shown that many dolphin and whale species are extraordinarily intelligent and social creatures, with complex cultures and rich inner lives. They are, in a word, persons.[13]

As we have seen, these sentiments are in many ways correct: dolphins do perform well on a number of tests looking at "human-like" cognitive abilities (i.e., intelligence), they display behavior that could be considered a product of culture, many species do live in complex societies, and it appears that they possess some form of self-awareness that could possibly be classified as resulting in "rich inner lives." But as this book has made clear, there is much more to the story than this. Science simply does not know enough about the dolphin mind, nor the minds of other animals, to justify the conclusion that dolphins are "extraordinarily intelligent." Depending on how one chooses to define the terms "intelligence," "culture," and "rich inner lives," these properties might well be widespread in the animal kingdom (and thus, by definition, not extraordinary). Intelligence itself is not even a valid scientific construct insofar as it's often shorthand for judging how human-like an animal's behavior is; consequently, there can be no research that "shows" this kind of subjective and unscientific assessment.

Producing advocacy statements concerning how we should be treating dolphins based on the current science of animal minds will always leave the door open for legitimate criticism from impartial scientists or critics who will be quick to note the current limitations of our knowledge. As Marian Stamp Dawkins noted in her discussion of how little science understands of animal consciousness, "going beyond what the scientific evidence actually allows us to say will inevitably lead to someone pointing this out."[14] Dawkins notes that over-interpreting or exaggerating scientific results when presenting ethical arguments, which often leads to criticism from many scientists, could "backfire in the end like 'crying wolf' once too often,"[15] and is therefore not the best means of making a watertight argument for the ethical treatment of animals. The basic tension here is between Dawkins' view that the science of animal consciousness is too murky to allow us to use it as the basis of a discussion concerning animal ethics, and the position put forth by Bekoff that "while the mystery of consciousness may remain and the knowledge about the nitty-gritty details of consciousness may still (and perhaps always) remain elusive, we know enough now to use it in interpretations and explanations of animal behavior and in arguments for animal protection."[16] Both positions offer an argument for the ethical treatment of animals, but disagree about the best way to interpret the current body of scientific literature concerning animal minds.

There is infinite wiggle room when it comes to interpreting the scientific evidence, which could allow for powerful arguments for the ethical treatment of animals to be presented under the heading of "when in doubt, err on the side of caution." But there is a real danger in this approach of oversimplifying or exaggerating the scientific findings, ignoring the scientific debates on these subjects, or downplaying the nature of ambiguous findings in order to uphold a succinct advocacy message. These tactics are sure to raise the hackles of those scientists whose very *raison d'être* hinges on describing and investigating the enigmatic nature of animal minds, animal behavior, and tackling the hard problem of consciousness. But there is also a real danger of throwing the baby out with the bathwater (once again)[17] by dismissing the evidence of what we do know about animal minds (e.g., dolphins have some form of self-awareness and metacognition) just because the study of consciousness or self-consciousness in animals is both difficult and prone to over-interpretation. Navigating this fine line (and properly demarcating where facts end and speculation begins) in order to produce the best argument for the ethical treatment of dolphins will require a healthy dialogue between advocates and their opponents, as well as contributions from impartial scientists who can remain detached and dispassionate when evaluating the scientific evidence that bears on this discussion. As an impartial scientist, I would not necessarily disagree with the above statement by *Wired Science*, but I would invoke Ben Goldacre's edict that "I think you'll find it's a bit more complicated than that."[18]

So what does a deconstructed dolphin tell us about how we should treat dolphins? Philosophers can argue about whether or not complex behavioral or cognitive properties are required for an animal to experience "harm" in the philosophical sense, but it's important to remember that science itself cannot make this judgment. Thus the naked results of scientific research into these phenomena do not—by themselves—constitute evidence for "extraordinariness," "worth," or "specialness" when discussing how we should treat dolphins. These notions only crop up when we begin comparing animals to humans, with the default assumption that human beings are the worthiest of all species as far as moral standing is concerned. This is the *de facto* basis for law in most countries, and it's therefore a worthwhile endeavor to make the human/animal comparison in a courtroom when one presents legal arguments for animal personhood. But let me be clear: arguing either for or against dolphin specialness is an exercise in philosophy or politics, but *not* science.

Consider an alternative approach to animal ethics raised by a number of philosophers[19,20] and scholars[21] (including dolphin scientists like Diana Reiss[22] and Fabienne Delfour[23]) that using the scientific evidence of dolphin cognitive complexity as arguments as to why dolphins deserve special treatment might not be the best approach to begin with. This viewpoint suggests that the tension we see between Dawkins' and Bekoff's approach to interpreting (and not over-interpreting) what the science of the animal mind has shown[24] can be avoided altogether if we refrain from framing the problem of animal ethics as some sort of contest to see which animals display enough complexity in terms of cognition, awareness, language skill, brain structure, culture, etc. to warrant special attention. Many animal rights and animal welfare advocates argue for increased protection based solely on minimizing or eliminating animal suffering (or enhancing animal pleasure), regardless of what the science might (or might not) be saying about behavioral or cognitive complexity. And those involved in environmental or animal conservation rarely appeal to the cognitive abilities of animals when presenting reasons for preserving species or ecosystems. So following this approach, the information found in this book could, while being vital to the scientific understanding of dolphins, have no bearing whatsoever on a discussion of how dolphins deserve to be treated.

If I may take off my scientist's hat for the moment and be allowed one final thought that is informed by my heart and not my head, it is this: I think it is time we stop judging (both explicitly and implicitly) dolphins and other animals based on whether their external behavior or their inner lives resembles that of human beings. I'm not making this plea as a legal, ethical, or philosophical argument, but as a general appeal to the intrinsic naturalist in all of us concerning how we should view humanity's place in the animal kingdom. Dolphins are marvelous, wondrous, and charismatic animals, but if we can stop looking at them through the narrow lens of the human condition by which we judge them as "special," we might open our minds and our hearts to the fact that many other species—from sharks to earwigs to rats—lead equally as wondrous and worthy lives.

ENDNOTES

Chapter 1

1. Russell, B. (1929). *Marriage and Morals.* George Allen and Unwin: London, 58.
2. D'Eath, R. B., & Keeling, L. J. (2003). Social discrimination and aggression by laying hens in large groups: from peck orders to social tolerance. *Applied Animal Behaviour Science, 84*(3), 197–212.
3. Sherwin, C. M., Heyes, C. M., & Nicol, C. J. (2002). Social learning influences the preferences of domestic hens for novel food. *Animal Behaviour 63*, 933–42.
4. Edgar, J. L., Lowe, J. C., Paul, E. S., & Nicol, C. J. (2011). Avian maternal response to chick distress. *Proceedings of the Royal Society B: Biological Sciences, 278*(1721), 3129–34.
5. Evans, C. S., & Evans, L. (1999). Chicken food calls are functionally referential. *Animal Behaviour, 58*, 307–19.
6. Evans, C. S., Evans, L., & Marler, P. (1993). On the meaning of alarm calls: functional reference in an avian vocal system. *Animal Behaviour, 46*, 23–38.
7. Abeyesinghe, S. M., Nicol, C. J., Hartnell, S. J., & Watges, C. M. (2005). Can domestic fowl, *Gallus gallus domesticus*, show self-control? *Animal Behaviour 70*, 1–11.
8. Finn, J. K., Tregenza, T., & Norman, M. D. (2009). Defensive tool use in a coconut-carrying octopus. *Current Biology, 19*(23), R1069–R1070.
9. A phrase often associated with dolphins. It appears, for example, in the following article: Morgan the orca arrives safely in Tenerife. (November 29, 2011). Retrieved from http://www.rnw.nl/english/bulletin/morgan-orca-arrives-safely-tenerife.
10. Goldacre, B. (2009). *Bad Science.* London: Harper Perennial, 100.
11. The Dolphins & Teleportation Symposium held at Kona, Hawaii from June 19–24, 2011, included discussions and activities involving "Time Travel to Mars" and dolphins teleporting people into the future. For more info on this see Ocean, J. (1997). *Dolphins into the Future.* Hawaii: Dolphin Connection, 174.
12. Many court cases have suggested that dolphins should have similar rights to human beings, for example: SeaWorld sued over "enslaved" killer whales. (February 7, 2012). *BBC News.* Retrieved from http://www.bbc.co.uk/news/world-us-canada-16920866.
13. A wonderfully detailed and well researched history of Lilly's work from which much of the narrative I present here originates can be found in Burnett, D. G. (2012). *The Sounding of the Whale.* Chicago: University of Chicago Press, and also Burnett, D. G. (2010). A mind in the water. *Orion Magazine.*
14. Lilly, J. C., & Miller, A. M. (1962). Operant conditioning of the bottlenose dolphin with electrical stimulation of the brain. *Journal of Comparative and Physiological Psychology, 55*, 73–9.

15. Lilly, J. C. (1958). Some considerations regarding basic mechanisms of positive and negative types of motivations. *American Journal of Psychiatry, 115*(6), 498–504.
16. Ubell, E. (June 19, 1958). Dolphins have very complicated nervous systems: may even "talk." *The Tuscaloosa News.*
17. As D. Graham Burnett noted in *The Sounding of the Whale* (p. 612), Carl Sagan and John Lilly (as well as other Order members) used to exchange cryptic messages with each other to simulate what it might be like to receive messages from alien civilizations.
18. Lilly, J. C. (1962). *Man and Dolphin.* London: Victor Gollancz.
19. As quoted from a review of the book *Man and Dolphin* by Margret C. Tavolga and William N. Tavolga, described in Burnett, D. G. (2012). *The Sounding of the Whale.* Chicago: University of Chicago Press, 590.
20. As quoted from a review of the book *Man and Dolphin* by J. W. Atz described in Burnett, D. G. (2012). *The Sounding of the Whale.* Chicago: University of Chicago Press, 590.
21. See Lilly, J. C. (1967). *The Mind of the Dolphin.* Garden City, NY: Doubleday, 274–5.
22. Lilly, J. C. (1987). *Communication Between Man and Dolphin: The Possibilities of Talking with Other Species.* New York: Julian Press. Pg. 1.
23. Greenberg, P. (January 6, 2012). How scientists came to love the whale. [Review of the book *The Sounding of the Whale*]. *New York Times.* Retrieved from http://www.nytimes.com/2012/01/08/books/review/the-sounding-of-the-whale-by-d-graham-burnett-book-review.html?pagewanted=all.
24. Lilly, J. (1962). *Man and Dolphin.* London: Victor Gollancz, 15.
25. Hooper, J. (1983, January). John Lilly: altered states. *Omni Magazine.*
26. Pryor, K., & Norris, K. S. (Eds.). (1991). *Dolphin Societies: Discoveries and Puzzles.* Berkeley: University of California Press, 199–225.
27. Fraser, J., Reiss, D., Boyle, P., Lemcke, K., Sickler, J., Elliot, E., Newman, B., & Gruber, S. (2006). Dolphins in popular literature and media. *Society and Animals 14*(4), 321–49. Pg. 327.
28. Fraser, J., Reiss, D., Boyle, P., Lemcke, K., Sickler, J., Elliot, E., Newman, B., & Gruber, S. (2006). Dolphins in popular literature and media. *Society and Animals, 14*(4), 321–49. Pg. 327.
29. Marino, L. (2011). Cetaceans and primates: convergence in intelligence and self-awareness. *Journal of Cosmology, 14*, 1063–79.
30. Lilly, J. C. (1962). *Man and Dolphin.* London: Victor Gollancz, 3.
31. Herzing, D. L., & White, T. (1999). Dolphins and the question of personhood. *Etica Animali, 9*(98), 64–84. Pgs. 79–80.
32. Viegas, J. (January 22, 2010). Dolphins: second-smartest animal? *Discovery News.* Retrieved from http://news.discovery.com/animals/dolphins-smarter-brain-function.html.
33. Dye, L. (February 24, 2010). Are dolphins also persons? *ABC News.* Retrieved from http://abcnews.go.com/Technology/AmazingAnimals/dolphins-animal-closest-intelligence-humans/story?id=9921886#.TztzmlGILs0.

34. Hoare, P. (February 24, 2012). After research reveals dolphins have extraordinary intellects and emotional IQs greater than ours, expert ask: should they be treated as humans? *Daily Mail.* Retrieved from http://www.dailymail.co.uk/news/article-2105703/Dolphin-expert-asks-should-treated-humans.html#ixzz1nIhSE5aB.
35. Keim, B. (July 19, 2012). New science emboldens long shot bid for dolphin, whale rights. *Wired Science.* Retrieved from http://www.wired.com/wiredscience/2012/07/cetacean-rights.
36. PETA sues SeaWorld for violating orcas' constitutional rights (October 25, 2011). Retrieved from http://www.peta.org/b/thepetafiles/archive/2011/10/25/peta-sues-seaworld-for-violating-orcas-constitutional-rights.aspx.
37. This includes Dr. Ingrid Visser of the Orca Research Trust in New Zealand.
38. This lawsuit was dismissed on February 8, 2012, with the ruling that the Thirteenth Amendment only applies to humans and/or persons.
39. PETA sues SeaWorld for violating orcas' constitutional rights (October 25, 2011). Retrieved from http://www.peta.org/b/thepetafiles/archive/2011/10/25/peta-sues-seaworld-for-violating-orcas-constitutional-rights.aspx.
40. Grimm, D. (2011). Are dolphins too smart for captivity? *Science, 332* (6029), 526–9.
41. Two special issues on why research with captive marine mammals is important: *International Journal of Comparative Psychology*, 2010, 23(3) and 23(4).
42. Rose, N. A., Parsons, E. C. M., & Garinato, R. (2009). The case against marine mammals in captivity (4th edn.). Washington, DC: The Humane Society of the United States and the World Society for the Protection of Animals.
43. Dolphins deserve same rights as humans, say scientists. (February 21, 2012). *BBC News.* Retrieved from http://www.bbc.co.uk/news/world-17116882.
44. Gould, S. J. (1981). *The Mismeasure of Man.* New York: W. W. Norton & Company.
45. Wasserman, E. A. (1993). Comparative cognition: beginning the second century of the study of animal intelligence. *Psychological Bulletin, 113*(2), 211–28. Pg. 212.
46. Hauser, M. (2000). *Wild Minds: What Animals Really Think.* New York: Henry Holt and Company. See Hauser's discussion of intelligence on pp. xviii and 257.
47. Quote from Rodney Brooks in Lewis, S. K., & Levin, D. (January 20, 2011). What is intelligence? *PBS.* Retrieved from http://www.pbs.org/wgbh/nova/body/what-is-intelligence.html.
48. The "intelligent" side of sheep. (November 7, 2001). *BBC News.* Retrieved from http://news.bbc.co.uk/2/hi/uk_news/wales/1643842.stm.
49. Justice Potter Stewart, concurring opinion in Jacobellis v. Ohio 378 U.S. 184 (1964), regarding possible obscenity in *The Lovers.*
50. Nakajima, S., Arimitsu, K., & Lattal, K. M. (2002). Estimation of animal intelligence by university students in Japan and the United States. *Anthrozoös, 15,* 194–205.
51. Pinker, S. (1997). *How the Mind Works.* London: Penguin Books, 62.
52. Romanes, G. J. (1882). *Animal Intelligence.* London: Kegan Paul Trench & Co., 16.
53. Cook, R. G., & Wasserman, E. A. (2006). Relational discrimination learning in pigeons. In E. A. Wasserman & T. R. Zentall (Eds.), *Comparative Cognition* (pp. 307–24). New York: Oxford University Press, 307.

54. Herman, L. M. (2006). Intelligence and rational behaviour in the bottlenosed dolphin. In S. Hurley & M. Nudds (Eds.), *Rational Animals?* (pp. 439–67). Oxford: Oxford University Press, 441.
55. Roth, G., & Dicke, U. (2005). Evolution of the brain and intelligence. *Trends in Cognitive Sciences, 9*(5), 250–7. Pg. 250.
56. Kennedy, J. S. (1992). *The New Anthropomorphism*. Cambridge: Cambridge University Press.
57. For a discussion on this topic see Chapter 3 of Dawkins, M. S. (2012). *Why Animals Matter: Animal Consciousness, Animal Welfare, and Human Well-being*. Oxford: Oxford University Press.
58. Dennett, D. C. (1987). *The Intentional Stance*. Cambridge, MA: MIT Press.
59. Sheehan, M. J., & Tibbetts, E. A. (2011). Specialized face learning is associated with individual recognition in paper wasps. *Science*, 334, 1272–5.
60. The "intelligent" side of sheep. (November 7, 2001). *BBC News*. Retrieved from http://news.bbc.co.uk/2/hi/uk_news/wales/1643842.stm.
61. Wynne, C. D. L. (2006). What are animals? Why anthropomorphism is still not a scientific approach to behavior. *Comparative Cognition Behavior Reviews, 2*, 125–35.
62. Doring, T. D., & Chittka L. (2011). How human are insects and does it matter? *Formosan Entomologist, 31*, 85–99.
63. Burghardt, G. M. (2006). Critical Anthropomorphism, Uncritical Anthropocentrism, and Naïve Nominalism. *Comparative Cognition Behavior Reviews, 2*, 136–8.
64. Budiansky, S. (1998). *If a Lion Could Talk: Animal Intelligence and the Evolution of Consciousness*. London: The Free Press, 3.
65. Consider the following statements in P. Brakes & M. P. Simmonds (Eds.) (2011). *Whales and Dolphins: Cognition, Culture, Conservation and Human Perceptions*. Oxford: Earthscan Publications. Compare the statement on p. 108: "It is certainly impossible to rank animal species by intelligence," with the statement on p. 2: "Our improved understanding of the lives of some of these animals reveals that they have some qualities shared with the primates, including ourselves, and this too arguably demonstrates that we need to pay them special attention." The argument here is that having primate-like intelligence (which presumably ranks higher than non-primate-like intelligence) is a reason for moral consideration.
66. Paul Manger quoted in Pidd, H. (September 11, 2006). Who's the dummy? *The Guardian*. Retrieved from http://www.guardian.co.uk/science/2006/sep/11/g2.
67. de Waal, F. (October 9, 2006). Looking at Flipper, seeing ourselves. *New York Times*. Retrieved from http://www.nytimes.com/2006/10/09/opinion/09dewaal.html.
68. Herman, L. M. (1980). Cognitive characteristics of dolphins. In L. M. Herman (Ed.), *Cetacean Behavior: Mechanisms and Function* (pp. 363–430). New York: Wiley Interscience.
69. Thorndike, E. L. (1911). *Animal Intelligence*. New York: Macmillan, 22.

Chapter 2

1. Wise, S. M. (2002). *Drawing the Line: Science and the Case for Animal Rights*. Cambridge, MA: Perseus Publishing, 133.

2. As late as 1983 Lilly stated that "I'm convinced that intelligence is a function of absolute brain size" in Hooper, J. (1983, January). John Lilly: altered states. *Omni Magazine*. For Lilly's original ideas on this subject, which I outline in this section, see Lilly, J. C. (1962). *Man and Dolphin*. London: Victor Gollancz.
3. Marino, L. (2002). Brain size evolution. In W. F. Perrin, B. Würsig, & J. G. M. Thewissen (Eds.), *Encyclopedia of Marine Mammals* (pp. 158–62). New York: Academic Press, 150.
4. Nelson, G. E. (1982). *Fundamental Concepts of Biology*. New York: Wiley, 262.
5. Shoshani, J., Kupsky, W. J., & Marchant, G. H. (2006). Elephant brain. Part I: Gross morphology functions, comparative anatomy, and evolution. *Brain Research Bulletin, 70*, 124–57. Pg. 124.
6. Marino, L. (2002). Brain size evolution. In W. F. Perrin, B. Würsig, & J. G. M. Thewissen (Eds.), *Encyclopedia of Marine Mammals* (pp. 158–62). New York: Academic Press.
7. Weights taken from Klinowska, M. (1992). Brains, behaviour and intelligence in cetaceans (whales, dolphins, and porpoises). In Ö. D. Jónsson (Ed.), *Whales and Ethics* (pp. 23–37). Reykjavik: Fisheries Research Institute, University of Iceland Press.
8. Weights taken from: Tartarelli, G., & Bisconti, M. (2006). Trajectories and constraints in brain evolution in primates and cetaceans. *Human Evolution, 21*(3–4), 275–87.
9. McDaniel, M. A. (2005). Big-brained people are smarter: a meta-analysis of the relationship between in vivo brain volume and intelligence. *Intelligence, 33*, 337–46.
10. For an overview, see Herculano-Houzel, S. (2009). The human brain in numbers: a linearly scaled-up primate brain. *Frontiers in Human Neuroscience, 3*, 31.
11. Healy, S. D., & Rowe, C. (2007). A critique of comparative studies of brain size. *Proceedings of the Royal Society B Biological Sciences, 274*(1609), 453–64.
12. See an overview of insect intelligence in Chittka, L. & Niven, J. (2009). Are bigger brains better? *Current Biology, 19*(21), R995–R1008.
13. Jerison, H. J. (1977). The theory of encephalization. *Annals of the New York Academy of Sciences, 299*, 146–60.
14. Jerison, H. J. (1977). The theory of encephalization. *Annals of the New York Academy of Sciences, 299*, 146–60.
15. Lori Marino quoted in Viegas, J. (January 22, 2010). Dolphins: second-smartest animal? *Discovery News*. Retrieved from http://news.discovery.com/animals/dolphins-smarter-brain-function.html.
16. Marino, L. (2002). Brain size evolution. In W. F. Perrin, B. Würsig, & J. G. M. Thewissen (Eds.), *Encyclopedia of Marine Mammals* (pp. 158–62). New York: Academic Press.
17. Marino, L. (2002). Convergence of complex cognitive abilities in cetaceans and primates. *Brain, Behavior and Evolution, 59*, 21–32.
18. White, T. (2007). *In Defense of Dolphins: The New Moral Frontier*. Malden, MA: Blackwell Publishing, 35.
19. Changizi, M. A. (2003). The relationship between number of muscles, behavioral repertoire size, and encephalization in mammals. *Journal of Theoretical Biology, 220*, 157–68.

20. Sol, D., Bacher, S., Reader, S. M., & Lefebvre, L. (2008). Brain size predicts the success of mammal species introduced into novel environments. *American Naturalist, 172*, S63–S71.
21. Clutton-Brock, T. H., & Harvey, P. H. (1980). Primates, brains and ecology. *Journal of Zoology, 190*(3), 309–23.
22. Deaner, R. O., Isler, K., Burkart, J., & Van Schaik, C. (2007). Overall brain size, and not encephalization quotient, best predicts cognitive ability across non-human primates. *Brain Behavior and Evolution, 70*(2), 115–24.
23. Gibson, K. R., Rumbaugh, D., & Beran, M. (2001). Bigger is better: primate brain size in relationship to cognition. In D. Falk & K. R. Gibson (Eds.), *Evolutionary Anatomy of the Primate Cerebral Cortex* (pp. 79–97). Cambridge: Cambridge University Press.
24. Marino, L. (1998). A comparison of encephalization between odontocete cetaceans and anthropoid primates. *Brain, Behavior and Evolution, 51*, 230–8.
25. Conner, R. C., Mann, J., Tyack, P. L., & Whitehead, H. (1998). Quantifying brain-behavior relations in cetaceans and primates—Reply. *Trends in Ecology & Evolution, 13*(10), 408.
26. Marino, L. (2004). Cetacean brain evolution: multiplication generates complexity. *International Journal of Comparative Psychology*, 17, 1–16.
27. Prior, H., Schwarz, A., & Güntürkün, O. (2008). Mirror-induced behavior in the magpie (*Pica pica*): evidence of self-recognition. *PLoS Biol* 6(8): e202. doi:10.1371/journal.pbio.0060202.
28. Pepperberg, I. M. (1998). Talking with Alex: logic and speech in parrots. *Scientific American, 9*(4), 60–5.
29. Shoshani, J., Kupsky, W. J., & Marchant, G. H. (2006). Elephant brain. Part I: Gross morphology, functions, comparative anatomy, and evolution. *Brain Research Bulletin, 70*(2), 124–57.
30. Roth, G., & Dicke, U. (2005). Evolution of the brain and intelligence. *Trends in Cognitive Sciences, 9*(5), 250–7.
31. Plotnik, J. M., de Waal, F. B. M., Moore, D., & Reiss, D. (2010). Self-recognition in the Asian elephant and future directions for cognitive research with elephants in zoological settings. *Zoo Biology, 29*(2), 179–91.
32. Rajala, A. Z., Reininger, K. R., Lancaster, K. M., & Populin, L. C. (2010). Rhesus monkeys (*Macaca mulatta*) do recognize themselves in the mirror: implications for the evolution of self-recognition. (J. Lauwereyns, Ed.) *PLoS ONE, 5*(9), 8.
33. Anderson, J. R., & Gallup, G. G., Jr. (2011). Do rhesus monkeys recognize themselves in mirrors? *American Journal of Primatology, 73*, 603–6.
34. Martin, R. D. (1984). Body size, brain size and feeding strategies. In D. J. Chivers, B. Wood, & A. Bilsborough (Eds.), *Food Acquisition and Processing in Primates* (pp. 73–103). New York: Plenum Press.
35. Anderson, J. R. (1983). Responses to mirror image stimulation and assessment of self-recognition in mirror- and peer-reared stumptail macaques. *The Quarterly Journal of Experimental Psychology B: Comparative and Physiological Psychology, 35*(3), 201–12.
36. Jerison, H. J. (1973). *Evolution of the Brain and Intelligence*. New York: Academic Press.

37. Fragaszy, D. M., Visalberghi, E., & Fedigan, L. M. (2004). *The Complete Capuchin: The Biology of the Genus Cebus*. Cambridge: Cambridge University Press.
38. Boddy, A. M., McGowen, M. R., Sherwood, C. C., Grossman, L. I., Goodman, M., & Wildman, D. E. (2012). Comparative analysis of encephalization in mammals reveals relaxed constraints on anthropoid primate and cetacean brain scaling. *Journal of Evolutionary Biology, 25*, 981–94.
39. See the discussion in the section Brain and Body Scaling: The Traditional View, in Herculano-Houzel, S. (2009). The human brain in numbers: a linearly scaled-up primate brain. *Frontiers in Human Neuroscience, 3*(November), 11. doi:10.3389/neuro.09.031.2009.
40. Boddy, A. M., McGowen, M. R., Sherwood, C. C., Grossman, L. I., Goodman, M., & Wildman, D. E. (2012). Comparative analysis of encephalization in mammals reveals relaxed constraints on anthropoid primate and cetacean brain scaling. *Journal of Evolutionary Biology, 25*, 981–94.
41. See the following for a discussion: Deaner, R. O., Isler, K., Burkart, J., & Van Schaik, C. (2007). Overall brain size, and not encephalization quotient, best predicts cognitive ability across non-human primates. *Brain Behavior and Evolution, 70*(2), 115–24.
42. Fichtelius, K. E,. Sjölander, S. (1972) *Smarter than Man? Intelligence in Whales, Dolphins, and Humans*. New York: Pantheon Books. Quote taken from back cover.
43. Ridgway, S. H. (1990). The central nervous system of the bottlenose dolphin. In S. Leatherwood & R. R. Reeves (Eds.), *The Bottlenose Dolphin* (pp. 69–97). New York: Academic Press, 75.
44. Oelschläger, H. H. A., & Oelschläger, J. S. (2009). Brain. In W. F. Perrin, B. Würsig, & J. G. M. Thewissen (Eds.), *Encyclopedia of Marine Mammals*, 2nd edn (pp. 134–49). New York: Academic Press.
45. See, for example, Pearson, H. C., & Shelton, D. E. (2010). A large-brained social animal. In B. Würsig & M. Würsig (Eds.), *The Dusky Dolphin: Master Acrobat off Different Shores* (pp. 333–53). San Diego, CA: Elsevier, Inc., 337.
46. Byrne, R. W., & Corp, N. (2004). Neocortex size predicts deception rate in primates. *Proceedings of the Royal Society B: Biological Sciences, 271*(1549), 1693–1699.
47. Lefebvre, L., Reader, S. M., & Sol, D. (2004). Brains, innovations and evolution in birds and primates. *Brain, Behavior and Evolution 63*, 233–46.
48. Dunbar, R. I. M. (1992). Neocortex size as a constraint on group size in primates. *Journal of Human Evolution, 20*, 469–93. doi:10.1016/0047-2484(92)90081-J.
49. Lefebvre, L., & Sol, D. (2008). Brains, lifestyles and cognition: are there general trends? *Brain, Behavior and Evolution, 72*, 135–44.
50. For a discussion see Pearson, H. C., & Shelton, D. E. (2010). A large-brained social animal. In B. Würsig & M. Würsig (Eds.), *The Dusky Dolphin: Master Acrobat off Different Shores* (pp. 333–53). San Diego, CA: Elsevier, Inc.
51. Connor, R. C. (2007). Dolphin social intelligence: complex alliance relationships in bottlenose dolphins and a consideration of selective environments for extreme brain size evolution in mammals. *Philosophical Transactions of the Royal Society B Biological Sciences, 362*(1480), 587–602.
52. Finarelli, J. A., & Flynn, J. J. (2009). Brain-size evolution and sociality in Carnivora. *Brain, 106*(23), 9345–9.

53. See Ridgway, S. H. (1990). The central nervous system of the bottlenose dolphin. In S. Leatherwood & R. R. Reeves (Eds.), *The Bottlenose Dolphin* (pp. 69–97). San Diego, CA: Academic Press, and Ridgway, S. H., & Au, W. W. L. (1999). Hearing and echolocation: dolphin. In G. Adelman & B. Smith (Eds.), *Elsevier's Encyclopedia of Neuroscience* (pp. 858–62). New York: Elsevier Science.
54. Marino, L. (2007). Cetacean brains: how aquatic are they? *Anatomical Record, 290*(6), 694–700.
55. Krubitzer, L., & Campi, K. (2009). Neocortical organization in monotremes. In L. R. Squire (Ed.), *Encyclopedia of Neuroscience* (pp. 51–9). Oxford: Academic Press.
56. Baron, G., Stepham, H., & Frahm, H. D. (1996). *Comparative Neurobiology in Chiroptera.* Berlin: Birkhauser-Verlag.
57. Hutcheon, J. M., Kirsch, J. A. W., & Garland, T., Jr. (2002). A comparative analysis of brain size in relation to foraging ecology and phylogeny in the chiroptera. *Brain Behavior and Evolution, 60,* 165–80.
58. Herculano-Houzel, S., Mota, B., & Lent, R. (2006). Cellular scaling rules for rodent brains. *Proceedings of the National Academy of Sciences,* 103, 12138–43.
59. Azevedo, F. A. C., Carvalho, L. R. B., Grinberg, L. T., Farfel, J. M., Ferretti, R. E. L., Leite, R. E. P., Jacob Filho, W., et al. (2009). Equal numbers of neuronal and nonneuronal cells make the human brain an isometrically scaled-up primate brain. *Journal of Comparative Neurology, 513*(5), 532–41.
60. Roth, G. (2000). The evolution and ontogeny of consciousness. In T. Metzinger (Ed.), *Neural Correlates of Consciousness: Empirical and Conceptual Questions* (pp. 77–97). Cambridge, MA: MIT Press.
61. Oelschläger, H. H. A., & Oelschläger, J. S. (2009). Brain. In W. F. Perrin, B. Würsig, & J. G. M. Thewissen (Eds.), *Encyclopedia of Marine Mammals,* 2nd edn (pp. 134–49). New York: Academic Press.
62. Huggenberger, S. (2008). The size and complexity of dolphin brains—a paradox? *Journal of the Marine Biological Association of the United Kingdom, 88*(06), 1103–8.
63. Roth, G., & Dicke, U. (2005). Evolution of the brain and intelligence. *Trends in Cognitive Sciences, 9*(5), 250–7.
64. Miklos, G. L. G. (1998). The evolution and modification of brains and sensory systems. *Daedalus, 127,* 197–216.
65. Oelschläger, H. H. A., & Oelschläger, J. S. (2009). Brain. In W. F. Perrin, B. Würsig, & J. G. M. Thewissen (Eds.), *Encyclopedia of Marine Mammals,* 2nd edn (pp. 134–49). New York: Academic Press.
66. Gourine, A. V., Kasymov, V., Marina, N., Tang, F., Figueiredo, M. F., Lane, S., Teschemacher, A. G., et al. (2010). Astrocytes control breathing through pH-dependent release of ATP. *Science, 329*(5991), 571–5.
67. Marino, L., Butti, C., Connor, R. C., Fordyce, R. E., Herman, L. M., Hof, P. R., Lefebvre, L., et al. (2008). A claim in search of evidence: reply to Manger's thermogenesis hypothesis of cetacean brain structure. *Biological Reviews of the Cambridge Philosophical Society, 83*(4), 417–40.
68. Marino, L., Connor, R. C., Fordyce, R. E., Herman, L. M., Hof, P. R., Lefebvre, L., Lusseau, D., et al. (2007). Cetaceans have complex brains for complex cognition. *PLoS Biology, 5*(5), e139.

69. Manger, P. R. (2006). An examination of cetacean brain structure with a novel hypothesis correlating thermogenesis to the evolution of a big brain. *Biological Reviews of the Cambridge Philosophical Society, 81*(2), 293–338.
70. Brownlow, M., Kvale, I., & Schofield, A. (2011). *Ocean Giants, Deep Thinkers* [Motion picture]. Season 1, Episode 2.
71. Chen, I. (2009, June). Brain cells for socializing. *Smithsonian Magazine.*
72. Lamm, C., & Singer, T. (2010). The role of anterior insular cortex in social emotions. *Brain Structure & Function, 241*(5–6), 579–951.
73. Butti, C., Santos, M., Uppal, N., Hof, P. R. (2011). Von Economo neurons: clinical and evolutionary perspectives. *Cortex.* doi:10.1016/j.cortex.2011.10.004.
74. Hof, P. R., & Van Der Gucht, E. (2007). Structure of the cerebral cortex of the humpback whale, *Megaptera novaeangliae* (Cetacea, Mysticeti, Balaenopteridae). *The Anatomical Record Part A Discoveries in Molecular Cellular and Evolutionary Biology, 31*, 1–31.
75. Nimchinsky, E. A., Gilisse, N. E., Allman, J. M., Perl, D. P., Erwin, J. M., & Hof, P. R. (1999). A neuronal morphologic type unique to humans and great apes. *Proceedings of the National Academy of Sciences of the United States of America,* 96(5), 268–73.
76. Hakeem, A. Y., Sherwood, C. C., Bonar, C. J., Butti, C., Hof, P. R., & Allman, J. M. (2009). Von Economo neurons in the elephant brain. *Anatomical Record, 292*(2), 242–8.
77. Hof, P. R., & Van Der Gucht, E. (2007). Structure of the cerebral cortex of the humpback whale, *Megaptera novaeangliae* (Cetacea, Mysticeti, Balaenopteridae). *The Anatomical Record Part A Discoveries in Molecular Cellular and Evolutionary Biology, 31*, 1–31.
78. Chen, I. (2009, June). Brain cells for socializing. *Smithsonian Magazine.*
79. Butti, C., Sherwood, C. C., Hakeem, A. Y., Allman, J. M., & Hof, P. R. (2009). Total number and volume of von Economo neurons in the cerebral cortex of cetaceans. *Journal of Comparative Neurology, 515*(2), 243–59.
80. Animals "can tell right from wrong": scientists suggest it's not just humans who have morals (May 26, 2009). *Daily Mail.* Retrieved from http://www.dailymail.co.uk/sciencetech/article-1187047/Animals-tell-right-wrong-Scientists-suggest-just-humans-morals.html.
81. Butti, C., Santos, M., Uppal, N., Hof, P. R. (2011). Von Economo neurons: clinical and evolutionary perspectives. *Cortex.* doi:10.1016/j.cortex.2011.10.004.
82. Seeley, W. W., Carlin, D. A., Allman, J. M., Macedo, M. N., Bush, C., Miller, B. L., & Dearmond, S. J. (2006). Early frontotemporal dementia targets neurons unique to apes and humans. *Annals of Neurology, 60*(6), 660–7.
83. Lori Marino stated the following: "The presence of these cells [von Economo neurons] is neurological support for the idea that cetaceans are capable of empathy and higher-order thinking and feeling," from this article: Whiting, C. C. (May 8, 2012). Humpback whales intervene in orca attack on gray whale calf. *Digital Journal Reports.* Retrieved from http://digitaljournal.com/article/324348.

84. Philippi, C. L., Feinstein, J. S., Khalsa, S. S., Damasio, A., Tranel, D., et al. (2012). Preserved self-awareness following extensive bilateral brain damage to the insula, anterior cingulate, and medial prefrontal cortices. *PLoS ONE, 7*(8), e38413. doi:10.1371/journal.pone.0038413.
85. Butti, C., Sherwood, C. C., Hakeem, A. Y., Allman, J. M., & Hof, P. R. (2009). Total number and volume of von Economo neurons in the cerebral cortex of cetaceans. *Journal of Comparative Neurology, 515*(2), 243–59.
86. Quote from Chen, I. (2009, June). Brain cells for socializing. *Smithsonian Magazine.*
87. Butti, C., Raghanti, M. A., Sherwood, C. C., & Hof, P. R. (2011). The neocortex of cetaceans: cytoarchitecture and comparison with other aquatic and terrestrial species. *Annals of the New York Academy of Sciences, 1225*(1), 47–58.
88. Evrard, H. C., Forro, T., & Logothetis, N. K. (2012). Von Economo neurons in the anterior insula of the macaque monkey. *Neuron, 74*(3), 482–9.
89. Marino, L. (2002). Brain size evolution. In W. F. Perrin, B. Würsig, & J. G. M. Thewissen (Eds.), *Encyclopedia of Marine Mammals* (pp. 158–62). New York: Academic Press, 150.
90. O'Shea, T. J., & Reep, R. L. (1990). Encephalization quotients and life-history traits in the Sirenia. *Journal of Mammalogy, 71*, 534–43. Pg. 534.
91. Van Essen, D. (1997). A tension-based theory of morphogenesis and compact wiring in the central nervous system. *Nature*, 385, 313–18.
92. Sherwood, C. C., Bauernfeind, A. L., Bianchi, S., Raghanti, M. A., & Hof, P. R. (2012). Human brain evolution writ large and small. (M. Hofman & D. Falk, Eds.). *Progress in Brain Research, 195*, 237–54.
93. Hof, P. R., & Van Der Gucht, E. (2007). Structure of the cerebral cortex of the humpback whale, *Megaptera novaeangliae* (Cetacea, Mysticeti, Balaenopteridae). *The Anatomical Record Part A Discoveries in Molecular Cellular and Evolutionary Biology, 31*, 1–31.
94. Hakeem, A. Y., Sherwood, C. C., Bonar, C. J., Butti, C., Hof, P. R., & Allman, J. M. (2009). Von Economo neurons in the elephant brain. *Anatomical Record, 292*(2), 242–8.
95. See, for example, Lori Marino's thoughts on this subject in Bekoff, M. (May 8, 2012). Humpback whales protect a gray whale from killer whales. *Psychology Today*. Retrieved from http://www.psychologytoday.com/blog/animal-emotions/201205/humpback-whales-protect-gray-whale-killer-whales.
96. Marino, L., Butti, C., Connor, R. C., Fordyce, R. E., Herman, L. M., Hof, P. R., Lefebvre, L., et al. (2008). A claim in search of evidence: reply to Manger's thermogenesis hypothesis of cetacean brain structure. *Biological Reviews of the Cambridge Philosophical Society, 83*(4), 417–40. Pg. 426.
97. de Waal, F. B. M., & Tyack, P. L. (2003). *Animal Social Complexity: Intelligence, Culture, and Individualized Societies.* Cambridge, MA: Harvard University Press.
98. Thanks to Dr. Patrick Hof for this insight.
99. Thanks to Dr. Patrick Hof for this explanation.
100. Manger, P. R. (2006). An examination of cetacean brain structure with a novel hypothesis correlating thermogenesis to the evolution of a big brain. *Biological Reviews of the Cambridge Philosophical Society, 81*(2), 293–338.

101. Semendeferi, K., Armostrong, E., Schleicher, A., Zilles, K., and Van Hoesen, G. W. (2001). Prefrontal cortex in humans and apes: a comparative study of area 10. *American Journal of Physical Anthropology, 114*, 224–241.
102. Maguire, E. A., Gadian, D. G., Johnsrude, I. S., Good, C. D., Ashburner, J., Frackowiak, R. S. J., & Frith, C. D. (2000). Navigation-related structural change in the hippocampi of taxi drivers. *Proceedings of the National Academy of Sciences of the United States of America, 97*(8), 4398–403.
103. Tamada, T., Miyauchi, S., Imamizu, H., Yoshioka, T., & Kawato, M. (1999). Cerebro-cerebellar functional connectivity revealed by the laterality index in tool-use learning. *NeuroReport, 10*(2), 325–31.
104. Finn, J. K., Tregenza, T., & Norman, M. D. (2009). Defensive tool use in a coconut-carrying octopus. *Current Biology, 19*(23) R1069–70.
105. Iwaniuk, A. N., Lefebvre, L., & Wylie, D. R. (2009). The comparative approach and brain-behaviour relationships: a tool for understanding tool use. *Canadian Journal of Experimental Psychology/Revue canadienne de psychologie expérimentale, 63*(2), 150–9.
106. Obayashi, S., Suhara, T., Kawabe, K., Okauchi, T., Maeda, J., Akine, Y., Onoe, H., & Iriki, A. (2001). Functional brain mapping of monkey tool use. *Neuroimage, 14*, 853–61.
107. Manger, P. R. (2006). An examination of cetacean brain structure with a novel hypothesis correlating thermogenesis to the evolution of a big brain. *Biological Reviews of the Cambridge Philosophical Society, 81*(2), 293–338.
108. Marino, L., Butti, C., Connor, R. C., Fordyce, R. E., Herman, L. M., Hof, P. R., Lefebvre, L., et al. (2008). A claim in search of evidence: reply to Manger's thermogenesis hypothesis of cetacean brain structure. *Biological Reviews of the Cambridge Philosophical Society, 83*(4), 417–40.
109. Marino, L., Connor, R. C., Fordyce, R., Herman, L. M., Hof, P. R., Lefebvre, L., et al. (2007). Cetaceans have complex brains for complex cognition. *PLoS Biology, 5*, 966–72. doi:10.1371/journal.pbio.0050139. e139.
110. Maximino, C. (2009). A quantitative test of the thermogenesis hypothesis of cetacean brain evolution, using phylogenetic comparative methods. *Marine and Freshwater Behaviour and Physiology, 42*(1), 1–17.
111. Manger quoted in Gregg, J. (September 5, 2006). The dim dolphin controversy. The dolphin pod. Retrieved from http://www.dolphincommunicationproject.org/index.php?option=com_content&task=view&id=1117&Itemid=285.
112. Marino, L., Butti, C., Connor, R. C., Fordyce, R. E., Herman, L. M., Hof, P. R., Lefebvre, L., et al. (2008). A claim in search of evidence: reply to Manger's thermogenesis hypothesis of cetacean brain structure. *Biological Reviews of the Cambridge Philosophical Society, 83*(4), 417–40.
113. Marino, L., Connor, R. C., Fordyce, R. E., Herman, L. M., Hof, P. R., Lefebvre, L., Lusseau, D., et al. (2007). Cetaceans have complex brains for complex cognition. *PLoS Biology, 5*(5), 7.
114. Consider the following quote from Manger: "Intelligence, measured by intelligent behaviour, is the observable expression . . . of neural activity and processing, and of course, if the processor is not built appropriately, as indicated in my

paper and the studies of many previously, then the actual ability for the production of intelligent behaviour must be missing. Without the appropriate brain, there is no intelligent behaviour." Quoted in Gregg, J. (September 5, 2006). The dim dolphin controversy. The dolphin pod. Retrieved from http://www.dolphincommunicationproject.org/index.php?option=com_content&task=view&id=1117&Itemid=285.

115. Quoted in Bonoguore, T. (March 17, 2009). Flipper no longer the head of the class? *Globe and Mail.* Retrieved from http://www.theglobeandmail.com/news/technology/science/article839604.ece.
116. Herman, L. M. (1980). Cognitive characteristics of dolphins. In L. M. Herman (Ed.), *Cetacean Behavior Mechanisms and Functions* (pp. 363–429). New York: Wiley Interscience, 363–4.

Chapter 3

1. Sherlock Holmes quoted in Conan Doyle, A. (1948). *The Adventure of the Blue Carbuncle.* New York: Baker Street Irregulars, 25.
2. Thomas White quoted in Connor, S. (February 21, 2012). Whales and dolphins are so intelligent they deserve same rights as humans, say experts. *The Independent.* Retrieved from http://www.independent.co.uk/environment/nature/whales-and-dolphins-are-so-intelligent-they-deserve-same-rights-as-humans-say-experts-7237448.html.
3. White, T. (2007). *In Defense of Dolphins: The New Moral Frontier.* Malden, MA: Blackwell Publishing, 80.
4. For a review of these topics see Bekoff, M., & Allen., C. (1997). Cognitive ethology: slayers, skeptics, and proponents. In R. W. Mitchell, N. S. Thompson, & H. L. Miles (Eds.), *Anthropomorphism, Anecdotes, and Animals* (pp. 313–34). Albany: State University of New York Press.
5. Longo, M. R., Schüür, F., Kammers, M. P. M., Tsakiris, M., & Haggard, P. (2008). What is embodiment? A psychometric approach. *Cognition, 107,* 978–98.
6. Herman, L. M., Matus, D. S., Herman, E. Y., Ivancic, M., & Pack, A. A. (2001). The bottlenosed dolphin's (*Tursiops truncatus*) understanding of gestures as symbol representations of its body parts. *Animal Learning & Behavior 29,* 250–64.
7. Herman, L. M. (2011). Body and self in dolphins. *Consciousness and Cognition, 21*(1), 526–45. Pg. 535.
8. Savage-Rumbaugh, E. S., Murphy, J., Sevcik, R. A., Brakke, K. E., Williams, S. L., & Rumbaugh, D. M. (1993). Language comprehension in ape and child. *Monographs for the Society for Research in Child Development, 58,* 1–221 [Serial No. 233].
9. Herman, L. M. (2002). Vocal, social, and self-imitation by bottlenosed dolphins. In K. Dautenhahn & C. Nehaniv (Eds.), Imitation in Animals and Artifacts (pp. 63–108). Cambridge, MA: MIT Press.
10. Mercado, E., Murray, S. O., Uyeyama, R. K., Pack, A. A., & Herman, L. M. (1998). Memory for recent actions in the bottlenosed dolphin (*Tursiops truncatus*): Repetition of arbitrary behaviors using an abstract rule. *Animal Learning Behavior, 26*(2), 210–18.

11. Zentall, T. R. (2008). Representing past and future events. In E. Dere, A. Easton, L. Nadel, & J. P. Huston (Eds.), *Handbook of Episodic Memory Research* (pp. 217–34). Oxford: Elsevier, 230.
12. Abramson, J. Z., Hernández-Lloreda, V., Call, J., & Colmenares, F. (2013). Experimental evidence for action imitation in killer whales (*Orcinus orca*). *Animal Cognition, 16*(1), 11–22.
13. Richards, D. G., Wolz, J. P., & Herman, L. M. (1984). Vocal mimicry of computer-generated sounds and vocal labeling of objects by a bottlenosed dolphin, *Tursiops truncatus. Journal of Comparative Psychology, 1*, 10–28.
14. Reiss, D., & McCowan, B. J. (1993). Spontaneous vocal mimicry and production by bottlenose dolphins (*Tursiops truncatus*): evidence for vocal learning. *Journal of Comparative Psychology, 107*(3), 301–12.
15. Janik, V. M. (2000). Whistle matching in wild bottlenose dolphins (Tursiops truncatus). *Science, 289*, 1355–57.
16. Ford, J. K. B. (1991). Vocal traditions among resident killer whales (*Orcinus orca*) in coastal waters of British Columbia. *Canadian Journal of Zoology/Revue Canadienne De Zoologie, 69*(6), 1454–83.
17. Foote, A. D., Griffin, R. M., Howitt, D., Larsson, L., Miller, P. J. O., & Rus Hoelzel, A. (2006). Killer whales are capable of vocal learning. *Biology Letters, 2*(4), 509–12.
18. Kremers, D., Jaramillo, M. B., Böye, M., Lemasson, A., & Hausberger, M. (2011). Do dolphins rehearse show-stimuli when at rest? Delayed matching of auditory memory. *Frontiers in Psychology, 2*(386) doi:10.3389/fpsyg.2011.00386.
19. May-Collado, L. J. (2010). Changes in whistle structure of two dolphin species during interspecific associations. *Ethology, 116*, 1065–74.
20. Herman, L. M., Morrel-Samuels, P., & Pack, A. A. (1990). Bottlenosed dolphin and human recognition of veridical and degraded video displays of an artificial gestural language. *Journal of Experimental Psychology: General, 119*(2), 215–30.
21. Bauer, G. B., & Johnson, C. M. (1994). Trained motor imitation by bottlenose dolphins (*Tursiops truncatus*). *Perceptual and Motor Skills, 79*, 1307–15.
22. Kuczaj, S. A., & Yeater, D. B. (2006). Dolphin imitation: who, what, when, and why? *Aquatic Mammals, 32*(4), 413–22.
23. Tomasello, M. (1996). Do apes ape? In C. M. Heyes & B. G. Galef, Jr. (Eds.), *Social Learning in Animals: The Roots of Culture* (pp. 319–46). New York: Academic Press.
24. Poole, J. H., Tyack, P. L., Stoeger-Horwath, A. S., and Watwood, S. (2005). Elephants are capable of vocal learning. *Nature* 434, 455–6.
25. Briefer, E., & McElligott, A. G. (2012). Social effects on vocal ontogeny in an ungulate, the goat (*Capra hircus*). *Animal Behaviour 83*, 991–1000.
26. Knörnschild, M., Nagy, M., Metz, M., Mayer, F., & von Helversen, O. (2010). Complex vocal imitation during ontogeny in a bat. *Biology Letters, 6*, 156–9.
27. Schusterman, R. J. (2008). Vocal learning in mammals with special emphasis on pinnipeds. In D. K. Oller & U. Gribel (Eds.), *The Evolution of Communicative Flexibility: Complexity, Creativity, and Adaptability in Human and Animal Communication* (pp. 41–70). Cambridge, MA: MIT Press.
28. Haun, D. B. M., & Call, J. (2008). Imitation recognition in great apes. *Current Biology, 18*(7), R288–R290.

29. Subiaul, F. (2007). The imitation faculty in monkeys: evaluating its features, distribution and evolution. *Journal of Anthropological Sciences, 85*, 35–62.
30. Gallup, G. G., Jr. (1970). Chimpanzees: self recognition. *Science, 167*(3914), 86–7.
31. de Waal, F. B. M., Dindo, M., Freeman, C. A., & Hall, M. (2005). The monkey in the mirror: hardly a stranger. *Proceedings of the National Academy of Sciences, 102*, 11140–7.
32. Gallup, G.G., Jr. (1970). Chimpanzees: self recognition. *Science, 167*(3914), 86–7. Pg. 87.
33. For an overview of these events see Reiss, D. (2011). *The Dolphin in the Mirror: Exploring Dolphin Minds and Saving Dolphin Lives*. Boston, MA: Houghton Mifflin Harcourt.
34. Marino, L., Reiss, D., & Gallup, G. (1993). Self-recognition in the bottlenose dolphin: a methodological test case for the study of extraterrestrial intelligence. In G. S. Shostak (Ed.), *Third Decennial US-USSR Conference on SETI. ASP Conference Series, 47*, 393.
35. For details on these kinds of behaviors see Sarko, D., Marino, L., & Reiss, D. (2003). A bottlenose dolphin's (*Tursiops truncatus*) responses to its mirror image: further analysis. *International Journal of Comparative Psychology, 15*, 69–76.
36. Thanks to Lori Marino for pointing this out; Pan and Delphi were apparently so agitated by sensing the mark on their skin that they would have been unable to remain calm enough to inspect themselves in the mirror.
37. Marten, K., & Psarakos, S. (1995). Evidence of self-awareness in the bottlenose dolphin (*Tursiops truncatus*). In S. T. Parker, R. Mitchell, & M. Boccia, *Self-Awareness in Animals and Humans: Developmental Perspectives* (pp. 361–79). Cambridge: Cambridge University Press.
38. Sarko, D., Marino, L., & Reiss, D. (2003). A bottlenose dolphin's (*Tursiops truncatus*) responses to its mirror image: further analysis. *International Journal of Comparative Psychology, 15*, 69–76. Pg. 70.
39. Anderson, J. R. (1995). Self-recognition in dolphins: credible cetaceans, compromised criteria, controls, and conclusions. *Consciousness and Cognition, 4*, 239–43.
40. Mitchell, R. W. (1995). Evidence of dolphin self-recognition and the difficulties of interpretation. *Consciousness and Cognition, 4*(2), 229–34.
41. Hart, D., & Whitlow, J. W., Jr. (1995). The experience of self in the bottlenose dolphin. *Consciousness and Cognition, 4*(2), 244–7.
42. Delfour, F., & Marten, K. (2001). Mirror image processing in three marine mammal species: killer whales (*Orcinus orca*), false killer whales (*Pseudorca crassidens*) and California sea lions (*Zalophus californianus*). *Behavioural Processes, 53*, 181–90.
43. Reiss, D., & Marino, L. (2001). Mirror self-recognition in the bottlenose dolphin: a case of cognitive convergence. *Proceedings of the National Academy of Sciences of the United States of America, 98*(10), 5937–42.
44. Reiss, D. (2011). *The Dolphin in the Mirror: Exploring Dolphin Minds and Saving Dolphin Lives*. Boston, MA: Houghton Mifflin Harcourt.
45. Reiss, D., & Marino, L. (2001). Mirror self-recognition in the bottlenose dolphin: a case of cognitive convergence. *Proceedings of the National Academy of Sciences of the United States of America, 98*(10), 5937–42. Pg. 5942.

46. Reiss, D., & Marino, L. (2001). Mirror self-recognition in the bottlenose dolphin: a case of cognitive convergence. *Proceedings of the National Academy of Sciences of the United States of America, 98*(10), 5937–42. Pg. 5937.
47. de Waal, F. B. M., Dindo, M., Freeman, C. A., & Hall, M. (2005). The monkey in the mirror: hardly a stranger. *Proceedings of the National Academy of Sciences, 102*, 11140–7.
48. Rajala, A. Z., Reininger, K. R., Lancaster, K. M., & Populin, L. C. (2010). Rhesus monkeys (*Macaca mulatta*) do recognize themselves in the mirror: implications for the evolution of self-recognition. *PLoS ONE, 5*(9), 8.
49. Note that the key difference between this study and other MSR studies is that the pigeons had been trained to use the mirror, as opposed to spontaneously using the mirror. Epstein, R., Lanza, R. P., & Skinner, B. F. (1981). "Self-awareness" in the pigeon. *Science, 212*, 695–6.
50. Ikeda, Y. (2009). A perspective on the study of cognition and sociality of cephalopod mollusks, a group of intelligent marine invertebrates. *Japanese Psychological Research, 51*(3), 146–53.
51. Delfour, F., & Herzing, D. (2009). Mirror exposure to free-ranging Atlantic spotted dolphins in the Bahamas. Presented at the 18th Biennial Conference on the Biology of Marine Mammals, Quebec City, Canada, October 2009.
52. Broesch, T., Callaghan, T., Henrich, J., Murphy, C., & Rochat, P. (2010). Cultural variations in children's mirror self-recognition. *Journal of CrossCultural Psychology, 42*(6), 1018–29.
53. Ledbetter, D. H., & Basen, J. D. (1982). Failure to demonstrate self-recognition in gorillas. *American Journal of Primatology, 2*, 307–10.
54. Patterson, F., & Gordon, W. (1993). The case for the personhood of gorillas. In P. Cavalieri & P. Singer (Eds.), *The Great Ape Project* (pp. 58–77). New York: St. Martin's Griffin.
55. Patterson, F. G. P., & Cohn, R. H. (1994). Self-recognition and self-awareness in lowland gorillas. In S. T. Parker, R. W. Mitchell, & M. L. Boccia (Eds.), *Self-Awareness in Animals and Humans* (pp. 273–90). New York: Cambridge University Press.
56. Bekoff, M. (2001). Observations of scent-marking and discriminating self from others by a domestic dog (*Canis familiaris*): tales of displaced yellow snow. *Behavioural Processes, 55*(2), 75–9. doi:10.1016/S0376-6357(01)00142-5.
57. Lockwood, J. A., & Rentz, D. C. F. (1996). Nest construction and recognition in a gryllacridid: the discovery of pheromonally mediated autorecognition in an insect. *Australian Journal of Zoology*, 44, 129–41.
58. Thanks to Lori Marino for this insight.
59. Mitchell, R. W. (1993). Mental models of mirror self-recognition: two theories. *New Ideas in Psychology, 11*, 295–325.
60. Mitchell, R. W. (1997). Kinesthetic-visual matching and the self-concept as explanations of mirror-self-recognition. *Journal for the Theory of Social Behavior, 27*(1), 17–39.
61. Heyes, C. M. (1994). Reflections on self-recognition in primates. *Animal Behaviour, 47*, 909–19.

62. Rochat, P., & Zahavi, D. (2011). The uncanny mirror: a reframing of mirror self-experience. *Consciousness and Cognition, 20,* 204–13.
63. Morin, A. (2011). Self-recognition, theory-of-mind, and self-awareness: what side are you on? *Laterality, 16*(3), 367–83.
64. de Waal, F. B. M., Dindo, M., Freeman, C. A., & Hall, M. (2005). The monkey in the mirror: hardly a stranger. *Proceedings of the National Academy of Sciences, 102,* 11140–7.
65. Carruthers, P. (2009). How we know our own minds: the relationship between mindreading and metacognition. *Behavioral and Brain Sciences, 32*(2), 121–38.
66. Humphrey, N. (1976). The social function of intellect. In P. P. G. Bateson & R. A. Hinde (Eds.), *Growing Points in Ethology* (pp. 303–17). Cambridge: Cambridge University Press.
67. For a review see Rochat, P., & Zahavi, D. (2011). The uncanny mirror: a reframing of mirror self-experience. *Consciousness and Cognition, 20,* 204–13.
68. Reiss, D., & Marino, L. (2001). Mirror self-recognition in the bottlenose dolphin: a case of cognitive convergence. *Proceedings of the National Academy of Sciences of the United States of America, 98*(10), 5937–42.
69. Marino, L. (2011). Brain structure and intelligence in cetaceans. In P. Brakes & M. Simmonds (Eds.), *Whales and Dolphins: Cognition, Culture, Conservation and Human Perceptions* (pp. 113–28). London: Earthscan, 125.
70. Statement of Lori Marino, PhD, Neuroscience and Behavioral Biology Program, Emory University, Atlanta, Georgia to The House Committee on Natural Resources Subcommittee on Insular Affairs, Oceans and Wildlife regarding educational aspects of public display of marine mammals, April 27, 2010.
71. Reiss, D. (2011). *The Dolphin in the Mirror: Exploring Dolphin Minds and Saving Dolphin Lives.* Boston, MA: Houghton Mifflin Harcourt, 167.
72. White, T. (2007). *In Defense of Dolphins: The New Moral Frontier.* Malden, MA: Blackwell Publishing, 65.
73. White, T. (2011). What is it like to be a dolphin? In P. Brakes & M. Simmonds (Eds.), *Whales and Dolphins: Cognition, Culture, Conservation and Human Perceptions* (pp. 188–206). London: Earthscan, 190.
74. Lemieux, L. (2009). *Rekindling the Waters: The Truth about Swimming with Dolphins.* Leicester: Matador, 287.
75. Frohoff, T. (2011). Lessons from dolphins. In P. Brakes & M. Simmonds (Eds.), *Whales and Dolphins: Cognition, Culture, Conservation and Human Perceptions* (pp. 135–9). London: Earthscan, 137.
76. For example: self-coherence, self-identity, self-experience, conceptualized self awareness, self-concept, cognitive self-consciousness, affective self-consciousness, auto-consciousness, situated consciousness, minimal phenomenal selfhood, reflective mind, sapience, perceptual consciousness, bi-directional consciousness, phenomenal consciousness, phenomenological consciousness, self-knowledge, psychological self-knowledge, private self-awareness, public self-awareness, meta self-awareness, etc.
77. Gallup, G. G., Jr. (1994). Self-recognition: research strategies and experimental design. In S. T. Parker, R. W. Mitchell, & M. L. Boccia (Eds.), *Self-Awareness in Animals*

and Humans: Developmental Perspectives (pp. 35–50). Cambridge: Cambridge University Press, 48.

78. Herman, L. M. (2011). Body and self in dolphins. *Consciousness and Cognition, 21*(1), 526–45.
79. White, T. (2007). *In Defense of Dolphins: The New Moral Frontier*. Malden, MA: Blackwell Publishing.
80. Herzing, D. L., & White, T. (1999). Dolphins and the question of personhood. *Etica Animali, 9*(98), 64–84.
81. Flavell, J. H. (1976). Metacognitive aspects of problem solving. In L. B. Resnick (Ed.), *The Nature of Intelligence* (pp. 231–5). Hillsdale, NJ: Lawrence Erlbaum Associates, 232.
82. Smith, J. D. (2012). Inaugurating the study of animal metacognition. *International Journal of Comparative Psychology, 23*(3), 401–13. Pg. 401.
83. Smith, J. D. (2012). Inaugurating the study of animal metacognition. *International Journal of Comparative Psychology, 23*(3), 401–13.
84. Smith, J. D., Schull, J., Strote, J., McGee, K., Egnor, R., & Erb, L. (1995). The uncertain response in the bottlenose dolphin (*Tursiops truncatus*). *Journal of Experimental Psychology, 124*, 391–408.
85. Carruthers, P. (2009). How we know our own minds: the relationship between mindreading and metacognition. *Behavioral and Brain Sciences, 32*(2), 121–38.
86. Smith, J. D., Coutinho, M. V. C., Boomer, J., & Beran, M. J. (2012). Metacognition across species. In J. Vonk & T. Shackelford (Eds.), *Oxford Handbook of Comparative Evolutionary Psychology* (pp. 271–96). Oxford: Oxford University Press, 285.
87. Browne, D. (2004). Do dolphins know their own minds? *Biology and Philosophy, 19*, 633–53.
88. Smith, J. D. (2012). Inaugurating the study of animal metacognition. *International Journal of Comparative Psychology, 23*(3), 401–13. Pg. 408.
89. See discussion on this by Browne, D. (2004). Do dolphins know their own minds? *Biology and Philosophy, 19*, 633–53.
90. Premack, D., & Woodruff, G. (1978). Does the chimpanzee have a theory of mind? *Behavioural and Brain Sciences, 1*, 515–26.
91. Herman, L. M. (2011). Body and self in dolphins. *Consciousness and Cognition, 21*(1), 526–45. Pg. 540.
92. Lurz, R. (2011). *Mindreading Animals: The Debate Over What Animals Know About Other Minds*. Cambridge, MA: MIT Press.
93. Pack, A. A., & Herman, L. M. (2004). Bottlenosed dolphins (*Tursiops truncatus*) comprehend the referent of both static and dynamic human gazing and pointing in an object-choice task. *Journal of Comparative Psychology, 118*(2), 160–71.
94. Tschudin, A., Call, J., Dunbar, R. I. M., Harris, G., & van der Elst, C. (2001). Comprehension of signs by dolphins (*Tursiops truncatus*). *Journal of Comparative Psychology, 115*, 100–5.
95. Flombaum, J. I., & Santos, L. R. (2005). Rhesus monkeys attribute perceptions to others. *Current Biology, 15*(5), 447–52.
96. Hare, B., Call, J., & Tomasello, M. (1998). Communication of food location between human and dog (*Canis familiaris*). *Evolution of Communication, 2*, 137–59.

97. Soproni, K., Miklosi, A., Topal, J., & Csanyi, V. (2001). Comprehension of human communicative signs in pet dogs (*Canis familiaris*). *Journal of Comparative Psychology, 115*(2), 122–6.
98. Kaminski, J., Riedel, J., Call, J., & Tomasello, M. (2005). Domestic goats, Capra hircus, follow gaze direction and use social cues in an object choice task. *Animal Behaviour, 69*, 11–18.
99. Wilkinson, A., Mandl, I., Bugnyar, T., & Huber, L. (2010). Gaze following in the red-footed tortoise (*Geochelone carbonaria*). *Animal Cognition, 13*, 765–9.
100. Moore, C., & Corkum, V. (1994). Social understanding at the end of the first year of life. *Developmental Review, 14*, 349–72.
101. Bugnyar, T., Stowe, M., & Heinrich, B. (2004). Ravens, *Corvus corax*, follow gaze direction of humans around obstacles. *Proceedings of the Royal Society B: Biological Sciences, 271*(1546), 1331–6.
102. Brauer, J., Call, J., & Tomasello, M. (2005). All great ape species follow gaze to distant locations and around barriers. *Journal of Comparative Psychology, 119*(2), 145–54.
103. See, for example, Povinelli, D. J., & Vonk, J. (2004). We don't need a microscope to explore the chimpanzee's mind. *Mind and Language, 19*(1), 1–28.
104. Shapiro, A. D., Janik, V. M., & Slater, P. J. (2003). A gray seal's (*Halichoerus grypus*) responses to experimenter-given pointing and directional cues. *Journal of Comparative Psychology, 117*(4), 355–62.
105. Neiworth, J. J., Burman, M. A., Basile, B. M., & Lickteig, M. T. (2002). Use of experimenter-given cues in visual co-orienting and in an object-choice task by a New World monkey species, cotton top tamarins (*Saguinus oedipus*). *Journal of Comparative Psychology, 116*(1), 3–11.
106. Inoue, Y., Inoue, E., & Itakura, S. (2004). Use of experimenter-given directional cues by a young white-handed gibbon (*Hylobates lar*). *Japanese Psychological Research, 46*(3), 262–7.
107. Anderson, J. R., Sallaberry, P., & Barbier, H. (1995). Use of experimenter-given cues during object-choice tasks by capuchin monkeys. *Animal Behaviour, 49*(1), 201–8.
108. Anderson, J. R., Montant, M., & Schmitt, D. (1996). Rhesus monkeys fail to use gaze direction as an experimenter-given cue in an object-choice task. *Behavioural Processes, 37*, 47–55.
109. Itakura, S. (1996). An exploratory study of gaze monitoring in nonhuman primates. *Japanese Psychological Research, 38*, 174–80.
110. Miklósi, Á., & Soproni, K. (2006). A comparative analysis of animals' understanding of the human pointing gesture. *Animal Cognition, 9*(2), 81–93.
111. Call, J., & Tomasello, M. (1994). Production and comprehension of referential pointing by orangutans (*Pongo pygmaeus*). *Journal of Comparative Psychology, 108*(4), 307–17.
112. Povinelli, D. J., Bering, J. M., & Giambrone, S. (2000). Toward a science of other minds: escaping the argument by analogy. *Cognitive Science, 24*(3), 509–41.
113. Itakura, S., & Tanaka, M. (1998). Use of experimenter-given cues during object-choice tasks by chimpanzees (*Pan troglodytes*), an orangutan (*Pongo pygmaeus*),

and human infants (*Homo sapiens*). *Journal of Comparative Psychology, 112*(2), 119–26.

114. Itakura, S., Agnetta, B., Hare, B., & Tomasello, M. (1999). Chimpanzee use of human and conspecific social cues to locate hidden food. *Developmental Science, 2*, 448–56.
115. Brauer, J., Kaminski, J., Riedel, J., Call, J., & Tomasello, M. (2006). Making inferences about the location of hidden food: social dog, causal ape. *Journal of Comparative Psychology, 120*(1), 38–47.
116. Hare, B., Brown, M., Williamson, C., & Tomasello, M. (2002). The domestication of social cognition in dogs. *Science, 298*(5598), 1634–6.
117. Miklósi, Á., Polgárdi, R., Topál, J., & Csányi, V. (1998). Use of experimenter-given cues in dogs. *Animal Cognition, 1*(2), 113–21.
118. Miklósi, Á., Pongracz, P., Lakatos, G., Topál, J., & Csányi, V. (2005). A comparative study of the use of visual communicative signals in interactions between dogs (*Canis familiaris*) and humans and cats (*Felis catus*) and humans. *Journal of Comparative Psychology, 119*(2), 179–86.
119. Soproni, K., Miklósi, A., Topál, J., & Csányi, V. (2001). Comprehension of human communicative signs in pet dogs (*Canis familiaris*). *Journal of Comparative Psychology, 115*(2), 122–6.
120. Agnetta, B., Hare, B., & Tomasello, M. (2000). Cues to food location that domestic dogs (*Canis familiaris*) of different ages do and do not use. *Animal Cognition, 3*(2), 107–12.
121. Kaminski, J., Riedel, J., Call, J., & Tomasello, M. (2005). Domestic goats, *Capra hircus*, follow gaze direction and use social cues in an object choice task. *Animal Behaviour, 69*, 11–18.
122. McKinley, J., & Sambrook, T. D. (2000). Use of human-given cues by domestic dogs (*Canis familiaris*) and horses (*Equus caballus*). *Animal Cognition, 3*(1), 13–22.
123. Herman, L., Pack, A. A., & Morrel-Samuels, P. (1993). Representational and conceptual skills of dolphins. In H. L. Roitblat, L. M. Herman, & P. E. Nachtigall (Eds.), *Language and Communication: Comparative Perspectives* (pp. 403–42). Hillsdale, NJ: Lawrence Erlbaum Associates.
124. Herman, L. M., Abichandani, S. L., Elhajj, A. N., Herman, E. Y. K., Sanchez, J. L., & Pack, A. A. (1999). Dolphins (*Tursiops truncatus*) comprehend the referential character of the human pointing gesture. *Journal of Comparative Psychology, 113*(4), 347–64.
125. Herman, L. M., & Uyeyama, R. U. (1999). The dolphin's grammatical competency: comments on Kako. *Animal Learning and Behavior 27*(1), 18–23.
126. Pack, A. A., & Herman, L. M. (2004). Bottlenosed dolphins (*Tursiops truncatus*) comprehend the referent of both static and dynamic human gazing and pointing in an object-choice task. *Journal of Comparative Psychology, 118*(2), 160–71.
127. Miklósi, Á., & Soproni, K. (2006). A comparative analysis of animals' understanding of the human pointing gesture. *Animal Cognition, 9*(2), 81–93.
128. Tschudin, A., Call, J., Dunbar, R. I., Harris, G., & van der Elst, C. (2001). Comprehension of signs by dolphins (*Tursiops truncatus*). *Journal of Comparative Psychology, 115*(1), 110–15.

129. Gregg, J. D., Dudzinski, K. M., & Smith, H. V. (2007). Do dolphins eavesdrop on the echolocation signals of conspecifics? *International Journal of Comparative Psychology, 20*, 65–88.
130. Xitco, M. J., Gory, J. D., & Kuczaj, S. A. I. (2001). Spontaneous pointing by bottlenose dolphins (*Tursiops trucatus*). *Animal Cognition, 4*, 115–23.
131. Xitco, M. J., Jr., Gory, J. D., & Kuczaj, S. A., II (2004). Dolphin pointing is linked to the attentional behavior of a receiver. *Animal Cognition, 8*, 231–8.
132. Dudzinski, K. M., Sakai, M., Masaki, K., Kogi, K., Hishii, T., & Kurimoto, M. (2003). Behavioural observations of bottlenose dolphins towards two dead conspecifics. *Aquatic Mammals, 29*(1), 108–16.
133. Leslie, A. M. (1994). ToMM, ToBY, and agency: core architecture and domain specificity. In L. A. Hirschfeld & S. A. Gelman (Eds.), *Mapping the Mind: Domain Specificity in Cognition and Culture* (pp. 119–48). New York: Cambridge University Press.
134. Baron-Cohen, S., Ring, H., Moriarty, J., Schmitz, B., Costa, D., & Ell, P. (1995). Recognition of mental state terms: clinical findings in children with autism, and a functional imaging study of normal adults. *British Journal of Psychiatry, 165*, 640–9.
135. Britton, N. F., Franks, N. R., Pratt, S. C., & Seeley, T. D. (2002). Deciding on a new home: how do honeybees agree? *Proceedings of the Royal Society B: Biological Sciences, 269*(1498), 1383–8.
136. Baron-Cohen, S. (1995). *Mindblindness: An Essay on Autism and Theory of Mind*. Cambridge, MA: MIT Press.
137. For example: Butterfill, S., & Apperly I. A. (In press). How to construct a minimal theory of mind. *Mind and Language, 28*.
138. See review in Apperly, I. A. (2012). What is "theory of mind"? Concepts, cognitive processes and individual differences. *Quarterly Journal of Experimental Psychology, 65*(5), 37–41.
139. For example: Butterfill, S., & Apperly I. A. (In press). How to construct a minimal theory of mind. *Mind and Language, 28*.
140. Astington, J. W., & Baird J. (2005). *Why Language Matters for Theory of Mind*. Oxford: Oxford University Press.
141. Butterfill, S., & Apperly I. A. (In press). How to construct a minimal theory of mind. *Mind and Language, 28*.
142. Lurz, R. (2011). *Mindreading Animals: The Debate over What Animals Know about Other Minds*. Cambridge, MA: MIT Press, 143.
143. Butterfill, S., & Apperly I. A. (In press). How to construct a minimal theory of mind. *Mind and Language, 28*.
144. Crockford, C., Wittig, R. M., Mundry, R., & Zuberbühler, K. (2011). Wild chimpanzees inform ignorant group members of danger. *Current Biology, 22*(2), 142–6. Pg. 145.
145. Call, J., & Tomasello, M. (2008). Does the chimpanzee have a theory of mind? 30 years later. *Trends in Cognitive Sciences, 12*(5), 187–92.
146. Hare, B., Call, J., & Tomasello, M. (2001). Do chimpanzees know what conspecifics know? *Animal Behaviour, 61*(1), 139–51.

147. Cheney, D. L. (2011). Extent and limits of cooperation in animals. *Proceedings of the National Academy of Sciences of the United States of America, 108*(Suppl. 2), 10902–9.
148. Wimmer, H., & Perner, J. (1983). Beliefs about beliefs: representation and constraining function of wrong beliefs in young children's understanding of deception. *Cognition, 13*(1), 103–28.
149. Call, J., & Tomasello, M. (2008). Does the chimpanzee have a theory of mind? 30 years later. *Trends in Cognitive Sciences, 12*(5), 187–92.
150. For a review see Tschudin, A. (2006). Belief attribution tasks with dolphins: what social minds reveal about animal rationality. In S. Hurley & M. Nudds (Eds.), *Rational Animals?* (pp. 411–36). Oxford: Oxford University Press.
151. Both of these objections were noted in Lurz, R. (2011). *Mindreading Animals: The Debate over What Animals Know about Other Minds.* Cambridge, MA: MIT Press, 151.
152. Tomonaga, M., Uwano, Y., Ogura, S., & Saito, T. (2010). Bottlenose dolphins' (*Tursiops truncatus*) theory of mind as demonstrated by responses to their trainers' attentional states. *International Journal of Comparative Psychology, 23*, 386–400.
153. Dally, J. M., Emery, N. J., & Clayton, N. S. (2004). Cache protection strategies by western scrub-jays (*Aphelocoma californica*): hiding food in the shade. *Proceedings of the Royal Society B: Biological Letters, 271*, 5387–90.
154. Clayton, N. S., Dally, J. M., & Emery, N. J. (2007). Social cognition by food-caching corvids: the western scrub-jay as a natural psychologist. *Philosophical Transactions of the Royal Society B: Biological Sciences, 362*(1480), 507–22.
155. Dally, J. M., Emery, N. J., & Clayton, N. S. (2006). Food-caching western scrub-jays keep track of who was watching when. *Science, 312*(5780), 1662–5.
156. Stulp, G., Emery, N. J., Verhulst, S., & Clayton, N. S. (2009). Western scrub-jays conceal auditory information when competitors can hear but cannot see. *Biology Letters, 5*(5), 583–5.
157. See discussion by Lurz, R. (2011). *Mindreading Animals: The Debate over What Animals Know about Other Minds.* Cambridge, MA: MIT Press, 55.
158. For an overview of the study of emotion in dolphins see Kuczaj, S. A., II, Highfill, L. E., Makecha, R. N., Byerly, H. C. (2012) Why do dolphins smile? A comparative perspective on dolphin emotions and emotional expressions. In S. Watanabe and S. A. Kuczaj (Eds.), *Emotions of Animals and Humans: Comparative Perspectives (The Science of the Mind)* (pp. 63–80). London: Springer.
159. Skinner, B. F. (1965). *Science and Human Behavior.* New York: Macmillan, 195.
160. Ekman, P. (1992). An argument for basic emotions. *Cognition & Emotion, 6*(3), 169–200.
161. Panksepp, J. (1998). *Affective Neuroscience: The Foundations of Human and Animal Emotions.* New York: Oxford University Press, 26.
162. Panksepp, J. (2011). Cross-species affective neuroscience decoding of the primal affective experiences of humans and related animals. *PLoS ONE, 6*(9), 15.
163. Prinz, J. (2004). *Gut Reactions: A Perceptual Theory of Emotions.* Oxford: Oxford University Press.

164. Winkielman, P., & C Berridge, K. (2004). Unconscious emotion. *Current Directions in Psychological Science, 13*(3), 120–3.
165. Berridge, K. C., & Winkielman, P. (2003). What is an unconscious emotion: the case for unconscious "liking." *Cognition and Emotion, 17*, 181–211.
166. Shewmon, D. A., Holmes, G. L., & Byrne, P. A. (1999). Consciousness in congenitally decorticate children: developmental vegetative state as self-fulfilling prophecy. *Developmental Medicine and Child Neurology, 41*(6), 364–74.
167. Balcombe, J. (2006). *Pleasurable Kingdom: Animals and the Nature of Feeling Good*. New York: Palgrave Macmillan.
168. Bekoff, M. (2007). *The Emotional Lives of Animals: A Leading Scientist Explores Animal Joy, Sorrow, and Empathy and Why They Matter*. Novato, CA: New World Library.
169. See, for example, Dawkins, M. S. (2012). *Why Animals Matter: Animal Consciousness, Animal Welfare, and Human Well-Being*. Oxford: Oxford University Press.
170. Bekoff, M. (ed.) (2000). *The Smile of a Dolphin: Remarkable Accounts of Animal Emotions*. London: Discovery Books.
171. Balcombe, J. (2006). *Pleasurable Kingdom: Animals and the Nature of Feeling Good*. New York: Palgrave Macmillan.
172. Herzing, D. A. (2000). Trail of grief. In M. Bekoff (Ed.), *The Smile of a Dolphin: Remarkable Accounts of Animal Emotions* (pp. 138–9). London: Discovery Books.
173. Rose, N. (2000). A death in the family. In M. Bekoff (Ed.), *The Smile of a Dolphin: Remarkable Accounts of Animal Emotions* (pp. 144–5). London: Discovery Books.
174. Frohoff, T. (2000). The dolphin's smile. In M. Bekoff (Ed.), *The Smile of a Dolphin: Remarkable Accounts of Animal Emotions* (pp. 78–9). London: Discovery Books.
175. Criticisms of this approach can be found in Dawkins, M. S. (2012). *Why Animals Matter: Animal Consciousness, Animal Welfare, and Human Well-Being*. Oxford: Oxford University Press.
176. Criticisms of this approach can be found in Sherwin, C. M. (2001). Can invertebrates suffer? Or, how robust is argument-by-analogy? *Animal Welfare*, 10 (Suppl. 1), S103–S118.
177. Criticisms of this approach can be found in Elwood, R. W. (2011). Pain and suffering in invertebrates? *ILAR Journal National Research Council Institute of Laboratory Animal Resources, 52*(2), 175–84.
178. Chalmers, D. J. (1995). Facing up to the problem of consciousness. *Journal of Consciousness Studies, 2*(3), 1–27.
179. Dawkins, M. S. (2001). Who needs consciousness? *Animal Welfare, 10*, S19–29. Pg. S28.
180. Panksepp, J. (2011). Cross-species affective neuroscience decoding of the primal affective experiences of humans and related animals. *PLoS ONE 6*(9), e21236. doi:10.1371/journal.pone.0021236.
181. Panksepp, J. (1992). A critical role for "affective neuroscience" in resolving what is basic about basic emotions. *Psychological Review, 99*(3), 554–60.
182. Panksepp, J., & Burgdorf, J. (2000). 50k-Hz chirping (laughter?) in response to conditioned and unconditioned tickle-induced reward in rats: effects of social housing and genetic variables. *Behavioral Brain Research, 115*, 25–38.
183. Panksepp, J. (2005). Affective consciousness: core emotional feelings in animals and humans. *Consciousness and Cognition, 14*, 30–80.

184. Low, P. (July 7, 2012). The Cambridge Declaration on Consciousness in Non-Human Animals. Signed at the Francis Crick Memorial Conference on Consciousness in Human and Non-Human Animals, at Churchill College, University of Cambridge.
185. Hauser, M. (2000). *Wild Minds: What Animals Really Think*. New York: Henry Holt and Company. See discussion of intelligence on pg. xviii.
186. Herzing, D. (2011). *Dolphins Diaries: My 20 Years with Spotted Dolphins in the Bahamas*. New York: St. Martin's Press, 106.
187. Herzing, D. L., & White, T. (1999). Dolphins and the question of personhood. *Etica Animali, 9*(98), 64–84.
188. Simmonds, M. (2006). Into the brains of whales. *Applied Animal Behaviour Science, 100*(1–2), 103–16.
189. Reiss, D. (2011). *The Dolphin in the Mirror: Exploring Dolphin Minds and Saving Dolphin Lives*. Boston, MA: Houghton Mifflin Harcourt, 202.
190. Kuczaj, S., Tranel, K., Trone, M., & Hill, H. (2001). Are animals capable of deception or empathy? Implications for animal consciousness and animal welfare. *Animal Welfare, 10*(1), S161–S173.
191. Wilson, E. O. (1975). *Sociobiology: The New Synthesis*. Cambridge, MA: Harvard University Press.
192. Dawkins, R. (1976). *The Selfish Gene*. New York: Oxford University Press.
193. Langford, D. J., Crager, S. E., Shehzad, Z., Smith, S. B., Sotocinal, S. G., Levenstadt, J. S., Chanda, M. L., et al. (2006). Social modulation of pain as evidence for empathy in mice. *Science, 312*(5782), 1967–70.
194. Edgar, J. L., Lowe, J. C., Paul, E. S., & Nicol, C. J. (2011). Avian maternal response to chick distress. *Proceedings of the Royal Society B: Biological Sciences, 278*(1721), 3129–34.
195. Singer, T., & Lamm, C. (2009). The social neuroscience of empathy. *Annals of the New York Academy of Sciences, 1156*(1), 81–96.
196. Preston, S. D., & de Waal, F. B. M. (2002). Empathy: its ultimate and proximate bases. *Behavioral and Brain Sciences, 25*(1), 1–20. Pg. 3.
197. Hatfield, E., Rapson, R. L., & Le, Y. L. (2007). Emotional contagion and empathy. In J. Decety and W. Ickes (Eds.), *The Social Neuroscience of Empathy*. Boston, MA: MIT Press.
198. Bartal, I. B. A., Decety, J., & Mason, P. (2011). Empathy and pro-social behavior in rats. *Science, 334*(6061), 1427–30.
199. Connor, R. C., & Norris, K. S. (1982). Are dolphins reciprocal altruists? *American Naturalist*, 119(3), 358–74.
200. Dudzinski, K. M., Gregg, J. D., Ribic, C. A., & Kuczaj, S. A. (2009). A comparison of pectoral fin contact between two different wild dolphin populations. *Behavioural Processes, 80*, 182–90.
201. Kuczaj, S., Tranel, K., Trone, M., & Hill, H. (2001). Are animals capable of deception or empathy? Implications for animal consciousness and animal welfare. *Animal Welfare, 10*(1), S161–S173.
202. de Waal, F. B. M. (2008). Putting the altruism back into altruism: the evolution of empathy. *Annual Review of Psychology, 59*(May 2007), 279–300.

203. Connor, R. C., & Norris, K. S. (1982). Are dolphins reciprocal altruists? *American Naturalist, 119*(3), 358–74.
204. Bartal, I. B. A., Decety, J., & Mason, P. (2011). Empathy and pro-social behavior in rats. *Science, 334*(6061), 1427–30.
205. For a discussion, see Preston, S. D., & de Waal, F. B. M. (2002). Empathy: its ultimate and proximate bases. *Behavioral and Brain Sciences, 25*(1), 1–20.
206. Like PETA: PETA sues SeaWorld for violating orcas' constitutional rights (October 25, 2011). Retrieved from http://www.peta.org/b/thepetafiles/archive/2011/10/25/peta-sues-seaworld-for-violating-orcas-constitutional-rights.aspx.
207. Preston, S. D., & de Waal, F. B. M. (2002). Empathy: its ultimate and proximate bases. *Behavioral and Brain Sciences, 25*(1), 1–20.
208. Schulte-Rüther, M., Markowitsch, H. J., Fink, G. R., & Piefke, M. (2007). Mirror neuron and theory of mind mechanisms involved in face-to-face interactions: a functional magnetic resonance imaging approach to empathy. *Journal of Cognitive Neuroscience, 19*(8), 1354–72.
209. Herzing, D. (2011). *Dolphins Diaries: My 20 Years with Spotted Dolphins in the Bahamas*. New York: St. Martin's Press, 270.
210. Hamilton, A. (2012). Reflecting on the mirror neuron system in autism: a systematic review of current theories. *Developmental Cognitive Neuroscience*. Retrieved from http://dx.doi.org/10.1016/j.dcn.2012.09.008.
211. Kilmer, J. M. (2011). More than one pathway to action understanding. *Trends in Cognitive Sciences, 15*(8), 352–7.
212. Frohoff, T. (2011). Lessons from dolphins. In P. Brakes & M. Simmonds (Eds.), *Whales and Dolphins: Cognition, Culture, Conservation and Human Perceptions* (pp. 135–8). London: Earthscan, 137.
213. Lori Marino stated the following: "The presence of these cells [von Economo neurons] is neurological support for the idea that cetaceans are capable of empathy and higher-order thinking and feeling," from this article: Whiting, C. C. (May 8, 2012). Humpback whales intervene in orca attack on gray whale calf. *Digital Journal Reports*. Retrieved from http://digitaljournal.com/article/324348.
214. Marino, L. (2004). Cetacean brain evolution: multiplication generates complexity. *International Journal of Comparative Psychology, 17*, 1–16.
215. Marino, L. (2011). Brain structure and intelligence in cetaceans. In P. Brakes & M. Simmonds (Eds.), *Whales and Dolphins: Cognition, Culture, Conservation and Human Perceptions* (pp. 113–28). London: Earthscan, 125.
216. Jerison, H. J. (1986). The perceptual worlds of dolphins. In R. J. Schusterman, J. Thomas, & F. G. Wood (Eds.), *Dolphin Cognition and Behavior: A Comparative Approach* (pp. 141–66). Hillsdale, NJ: Erlbaum.
217. Herzing, D. L., & White, T. (1999). Dolphins and the question of personhood. *Etica Animali, 9*(98), 64–84. Pg. 74.
218. White, T. (2007). *In Defense of Dolphins: The New Moral Frontier*. Malden, MA: Blackwell Publishing, 42.
219. Kuczaj, S., Tranel, K., Trone, M., & Hill, H. (2001). Are animals capable of deception or empathy? Implications for animal consciousness and animal welfare. *Animal Welfare, 10*(1), S161–S173.

220. Reiss, D. (2011). *The Dolphin in the Mirror: Exploring Dolphin Minds and Saving Dolphin Lives*. Boston, MA: Houghton Mifflin Harcourt, 246.
221. de Waal, F. (2009). *The Age of Empathy: Nature' s Lessons for a Kinder Society*. New York: Crown Publishers.

Chapter 4

1. Stan Kuczaj quoted in Grimm D. (2011). Are dolphins too smart for captivity? *Science*, 332(6029), 526–9. Pg. 528.
2. Lyn, H. (2012) Apes and the evolution of language: taking stock of 40 years of research. In J. Vonk & T. Shackelford (Eds.), *Oxford Handbook of Comparative Evolutionary Psychology* (pp. 356–80). Oxford: Oxford University Press.
3. Gardner, R. A., & Gardner, B. T. (1969). Teaching sign language to a chimpanzee. *Science, 165*(894), 664–72.
4. Herman, L. M., Richards, D. G., & Wolz, J. P. (1984). Comprehension of sentences by bottlenosed dolphins. *Cognition, 16*, 129–219.
5. Herman, L. M. (2009). Language learning and cognitive skills. In W. F. Perrin, B. Würsig, & H. C. M. Thewissen (Eds.), *Encyclopedia of Marine Mammals*, 2nd edn (pp. 657–63). New York Academic Press.
6. Fodor, J. A. (1975). *The Language of Thought*. Trowbridge, UK: Crowell Press.
7. See Chapter 3 of Pinker, S. (1994). *The Language Instinct*. New York: William Morrow.
8. Herman, L. M. (1986). Cognition and language competencies of bottlenosed dolphins. In R. J. Schusterman, J. Thomas, and F. G. Wood (Eds.), *Dolphin Cognition and Behavior: A Comparative Approach* (pp. 221–51). Hillsdale, NJ: Lawrence Erlbaum Associates.
9. Herman, L. M., Richards, D. G., & Wolz, J. P. (1984). Comprehension of sentences by bottlenosed dolphins. *Cognition, 16*, 129–219.
10. Herman, L. M., & Forestell, P. H. (1985). Reporting presence or absence of named objects by a language-trained dolphin. *Neuroscience and Behavioral Reviews*, 9, 667–91.
11. Herman, L. M., Kuczaj, S. A., II, & Holder, M. D. (1993). Responses to anomalous gestural sequences by a language-trained dolphin: evidence for processing of semantic relations and syntactic information. *Journal of Experimental Psychology: General, 122*, 184–94. Pg. 185.
12. Herman, L. M. (2006). Intelligence and rational behaviour in the bottlenosed dolphin. In S. Hurley & M. Nudds (Eds.), *Rational Animals?* (pp. 439–67). Oxford: Oxford University Press, 443.
13. Pack, A. A. (2010). The synergy of laboratory and field studies of dolphin behavior and cognition. *International Journal of Comparative Psychology, 23*, 538–65.
14. Herman, L. M., Richards, D. G., & Wolz, J. P. (1984). Comprehension of sentences by bottlenosed dolphins. *Cognition, 16*, 129–219.
15. Herman, L. M. (1987). Receptive competences of language-trained animals. In J. S. Rosenblatt, C. Beer, M. C. Busnel, & P. J. B. Slater (Eds.), *Advances in the Study of Behavior*, Vol. 17 (pp. 1–60). Petaluma, CA: Academic Press.

16. Herman, L. M., Kuczaj, S. A., & Holder, M. D. (1993). Responses to anomalous gestural sequences by a language-trained dolphin: evidence for processing of semantic relations and syntactic information. *Journal of Experimental Psychology: General, 122*(2), 184–94.
17. Gallistel, C. R. (1998). Symbolic processes in the brain: the case of insect navigation. In D. Scarorough & S. Sternberg (Eds.), *Methods, Models and Conceptual Issues.* Vol. 4: An Invitation to Cognitive Science, 2nd edn (pp. 1–51). Cambridge, MA: MIT Press.
18. Cruse, H., & Wehner, R. (2011). No need for a cognitive map: decentralized memory for insect navigation. *PLoS Computational Biology, 7*(3), 10.
19. Herrnstein, R. J., & Loveland, D. H. (1964). Complex visual concept in the pigeon. *Science, 146*(3643), 549–50.
20. Savage-Rumbaugh, E. S., Rumbaugh, D. M., Smith, S. T., & Lawson, J. (1980). Reference: the linguistic essential. *Science, 210*(4472), 922–5.
21. Chittka, L., & Jensen, K. (2011). Animal cognition: concepts from apes to bees. *Current Biology, 21,* R116–R119.
22. Cook, R. G., Katz, J. S., & Cavoto, B. R. (1997). Pigeon same-different concept learning with multiple stimulus classes. *Journal of Experimental Psychology: Animal Behavior Processes, 23*(4), 417–33.
23. Blaisdell, A. P., & Cook, R. G. (2005). Two-item same-different concept learning in pigeons. *Learning Behavior: A Psychonomic Society Publication, 33*(1), 67–77.
24. Browne, D. (2004). Do dolphins know their own minds? *Biology and Philosophy 19,* 633–53.
25. Perner, J. (1991). *Understanding the Representational Mind: Learning Development and Conceptual Change.* Cambridge, MA: MIT Press.
26. Herman, L. M., Morrel-Samuels, P., & Pack, A. A. (1990). Bottlenosed dolphin and human recognition of veridical and degraded video displays of an artificial gestural language. *Journal of Experimental Psychology: General, 119*(2), 215–30.
27. Savage-Rumbaugh, E. S. (1986). *Ape Language: From Conditioned Response to Symbol.* New York: Columbia University Press.
28. Herman, L. M. (2006). Intelligence and rational behaviour in the bottlenosed dolphin. In S. Hurley & M. Nudds (Eds.), *Rational Animals?* (pp. 439–67). Oxford: Oxford University Press, 448.
29. Herman, L. M. (2006). Intelligence and rational behaviour in the bottlenosed dolphin. In S. Hurley & M. Nudds (Eds.), *Rational Animals?* (pp. 439–67). Oxford: Oxford University Press, 449.
30. Jaakola, K. (2012). Cetacean cognitive specializations. In J. Vonk & T. Shackelford (Eds.), *Oxford Handbook of Comparative Evolutionary Psychology* (pp. 144–65). Oxford: Oxford University Press, 155.
31. Menzel, E. W., Savage-Rumbaugh, E. S., & Lawson, J. (1985). Chimpanzee (*Pan troglodytes*) spatial problem solving with the use of mirrors and televised equivalents of mirrors. *Journal of Comparative Psychology, 99*(2), 211–17.
32. Keeling, L. J., & Hurnik, J. F. D. A. A. (1993). Chickens show socially facilitated feeding behaviour in response to a video image of a conspecific. *Applied Animal Behaviour Science, 36,* 223–31.

33. Clark, D. L., & Uetz, G. W. (1990). Video image recognition by the jumping spider, *Maevia inclemens* (Araneae: Salticidae). *Animal Behaviour, 40*(5), 884–90.
34. Herman, L. M., Uyeyama, R. K., & Pack, A. A. (2008). Bottlenose dolphins understand relationships between concepts. *Behavioral and Brain Sciences, 31*, 139–40.
35. Mercado, E., Killebrew, D., Pack, A., Mácha, I., & Herman, L. (2000). Generalization of "same-different" classification abilities in bottlenosed dolphins. *Behavioural Processes, 50*(2–3), 79–94.
36. Jaakkola, K., Fellner, W., Erb, L., Rodriguez, M., & Guarino, E. (2005). Understanding of the concept of numerically "less" by bottlenose dolphins (*Tursiops truncatus*). *Journal of Comparative Psychology, 119*(3), 296–303.
37. Kilian, A., Yaman, S., Von Fersen, L., & Güntürkün, O. (2003). A bottlenose dolphin discriminates visual stimuli differing in numerosity. *Learning Behavior: a Psychonomic Society Publication, 31*(2), 133–42.
38. Murayama, T., Usui, A., Takeda, E., Kato, K., & Maejima, K. (2012). Relative size discrimination and perception of the Ebbinghaus illusion in a bottlenose dolphin (*Tursiops truncatus*). *Aquatic Mammals, 38*(4), 333–42.
39. Herman, L. M., Pack, A. A., & Morrel-Samuels, P. (1993). Representational and conceptual skills of dolphins. In H. L. Roitblat, L. M. Herman, & P. E. Nachtigall (Eds.), *Language and Communication: Comparative Perspectives* (pp. 403–42). Mahwah, NJ: Lawrence Erlbaum Associates.
40. Ralston, J. V., & Herman, L. M. (1995). Perception and generalization of frequency contours by a bottlenose dolphin (*Tursiops truncatus*). *Journal of Comparative Psychology, 109*, 268–77.
41. Harley, H. (2008). Whistle discrimination and categorization by the Atlantic bottlenose dolphin (*Tursiops truncatus*): a review of the signature whistle framework and a perceptual test. *Behavioural Processes, 77*(2), 243–68.
42. Janik, V. M., Sayigh, L. S., & Wells, R. S. (2006). Signature whistle shape conveys identity information to bottlenose dolphins. *Proceedings of the National Academy of Sciences of the United States of America, 103*(21), 8293–7.
43. Jaakkola, K., Guarino, E., Rodriguez, M., Erb, L., & Trone, M. (2010). What do dolphins (*Tursiops truncatus*) understand about hidden objects? *Animal Cognition, 13*(1), 103–20.
44. Herman, L. M., Hovancik, J. R., Gory, J. D., & Bradshaw, G. L. (1989). Generalization of visual matching by a bottlenosed dolphin (*Tursiops truncatus*): evidence for invariance of cognitive performance with visual and auditory materials. *Journal of Experimental Psychology: Animal Behavior Processes, 15*(2), 124–36.
45. Harley, H. E., Putman, E. A., & Roitblat, H. L. (2003). Bottlenose dolphins perceive object features through echolocation. *Nature*, 424, 667–9.
46. Pack, A. A., Herman, L. M., Hoffmann-Kuhnt, M., & Branstetter, B. K. (2002). The object behind the echo: dolphins (*Tursiops truncatus*) perceive object shape globally through echolocation. *Behavioural Processes, 58*(1–2), 1–26.
47. Pack, A. A., & Herman, L. M. (1995). Sensory integration in the bottlenosed dolphin: immediate recognition of complex shapes across the senses of echolocation and vision. *Journal of the Acoustical Society of America, 98*, 722–33.

48. Pack, A. A., & Herman, L. (1996). Dolphins can immediately recognize complex shapes across the senses of echolocation and vision. *Journal of the Acoustical Society of America, 100*(4), 2610.
49. Harley, H. E., Roitblat, H. L., & Nachtigall, P. E. (1996). Object representation in the bottlenose dolphin (*Tursiops truncatus*): integration of visual and echoic information. *Journal of Experimental Psychology: Animal Behavior Processes, 22*(2), 164–74.
50. Herman, L. M. (2011). Body and self in dolphins. *Consciousness and Cognition, 21*(1), 526–45.
51. White, T. (2007). *In Defense of Dolphins: The New Moral Frontier*. Malden, MA: Blackwell Publishing, 39–40.
52. Shimojo, S., & Shams, L. (2001). Sensory modalities are not separate modalities: plasticity and interactions. *Current Opinion in Neurobiology, 11*(4), 505–9.
53. Elliott, R. C. (1977). Cross-modal recognition in three primates. *Neuropsychologia, 15*(1), 183–6.
54. Taylor, A. M., Reby, D., & McComb, K. (2011). Cross modal perception of body size in domestic dogs (*Canis familiaris*). *PLoS ONE, 6*(2), 6.
55. Winters, B. D., & Reid, J. M. (2010). A distributed cortical representation underlies crossmodal object recognition in rats. *Journal of Neuroscience, 30*(18), 6253–61.
56. Benoit, M. M., Raij, T., Lin, F.-H., Jääskeläinen, I. P., & Stufflebeam, S. (2010). Primary and multisensory cortical activity is correlated with audiovisual percepts. *Human Brain Mapping, 31*(4), 526–38.
57. Ward, J., & Simner, J. (2003). Lexical-gustatory synaesthesia: linguistic and conceptual factors. *Cognition*, 89, 237–61.
58. Newell, F. N., Ernst, M. O., Tjan, B. S., & Bülthoff, H. H. (2001). Viewpoint dependence in visual and haptic object recognition. *Psychological Science, 12*(1), 37–42.
59. Abramson, J. Z., Hernández-Lloreda, V., Call, J., & Colmenares, F. (2011). Relative quantity judgments in South American sea lions (*Otaria flavescens*). *Animal Cognition, 14*(5), 695–706.
60. Irie-Sugimoto, N., Kobayashi, T., Sato, T., & Hasegawa, T. (2009). Relative quantity judgment by Asian elephants (*Elephas maximus*). *Animal Cognition, 12*(1), 193–9.
61. Vonk, J., & Beran, M. J. (2012). Bears "count" too: quantity estimation and comparison in black bears, *Ursus americanus*. *Animal Behaviour, 84*(1), 231–8.
62. Thomas, R. K., & Chase, L. (1980). Relative numerousness judgments by squirrel monkeys. *Bulletin of the Psychonomic Society, 16*(2), 79–82.
63. Agrillo, C., Piffer, L., & Bisazza, A. (2010). Large number discrimination by mosquitofish. (G. Chapouthier, Ed.) *PLoS ONE, 5*(12), 10.
64. Uller, C., Jaeger, R., Guidry, G., & Martin, C. (2003). Salamanders (*Plethodon cinereus*) go for more: rudiments of number in an amphibian. *Animal Cognition, 6*(2), 105–12.
65. Dacke, M., & Srinivasan, M. V. (2008). Evidence for counting in insects. *Animal Cognition, 11*(4), 683–9.
66. Katz, J. S., & Wright, A. A. (2006). Same/different abstract-concept learning by pigeons. *Journal of Experimental Psychology: Animal Behavior Processes, 32*(1), 80–6.

67. Cook, R. G., & Brooks, D. I. (2009). Generalized auditory same-different discrimination by pigeons. *Journal of Experimental Psychology: Animal Behavior Processes, 35*(1), 108–15.
68. Bhatt, R. S., Wasserman, E. A., Reynolds, W. F., & Knauss, K. S. (1988). Conceptual behavior in pigeons: categorization of both familiar and novel examples from four classes of natural and artificial stimuli. *Journal of Experimental Psychology: Animal Behavior Processes, 14,* 219–34.
69. Watanabe, S. (2001). Van Gogh, Chagall and pigeons: picture discrimination in pigeons and humans. *Animal Cognition,* 4(3–4), 147–51.
70. Watanabe, S. (2010). Pigeons can discriminate "good" and "bad" paintings by children. *Animal Cognition, 13*(1), 75–85.
71. Jaakkola, K., Guarino, E., Rodriguez, M., Erb, L., & Trone, M. (2010). What do dolphins (*Tursiops truncatus*) understand about hidden objects? *Animal Cognition, 13*(1), 103–20.
72. Barth, J., & Call, J. (2006). Tracking the displacement of objects: a series of tasks with great apes (*Pan troglodytes, Pan paniscus, Gorilla gorilla,* and *Pongo pygmaeus*) and young children (*Homo sapiens*). *Journal of Experimental Psychology: Animal Behavior Processes, 32*(3), 239–52.
73. Collier-Baker, E., Davis, J. M., & Suddendorf, T. (2004). Do dogs (*Canis familiaris*) understand invisible displacement? *Journal of Comparative Psychology,* 118, 421–33.
74. Emery, N. J. (2006). Cognitive ornithology: the evolution of avian intelligence. *Philosophical Transactions of the Royal Society of London—Series B: Biological Sciences, 361*(1465), 23–43.
75. Light, K. R., Kolata, S., Wass, C., Denman-Brice, A., Zagalsky, R., & Matzel, L. D. (2010). Working memory training promotes general cognitive abilities in genetically heterogeneous mice. *Current Biology, 20*(8), 777–82.
76. Shipstead, Z., Redick, T. S., & Engle, R. W. (2012). Is working memory training effective? *Psychological Bulletin, 138*(4), 1–27.
77. For an overview see Jaakola, K. (2012). Cetacean cognitive specializations. In J. Vonk & T. Shackelford (Eds.), *Oxford Handbook of Comparative Evolutionary Psychology* (pp. 144–65). Oxford: Oxford University Press, 148.
78. Thompson, R. K. R., & Herman, L. M. (1977). Memory for lists of sounds by the bottlenosed dolphin: convergence of memory processes with humans? *Science, 195,* 501–3.
79. Merritt, D., Maclean, E. L., Jaffe, S., & Brannon, E. M. (2007). A comparative analysis of serial ordering in ring-tailed lemurs (*Lemur catta*). *Journal of Comparative Psychology, 121*(4), 363–71.
80. Herzing, D. L., & White, T. (1999). Dolphins and the question of personhood. *Etica Animali,* 9(98), 64–84. Pg. 75.
81. Fagot, J., & Cook, R. G. (2006). Evidence for large long-term memory capacities in baboons and pigeons and its implications for learning and the evolution of cognition. *Proceedings of the National Academy of Sciences of the United States of America, 103*(46), 17564–7.

82. Raby, C. R., & Clayton, N. S. (2012). Episodic memory and future planning. In J. Vonk & T. Shackelford (Eds.), *Oxford Handbook of Comparative Evolutionary Psychology* (pp. 217–35). Oxford: Oxford University Press, 217.
83. Tulving, E. (1983). *Elements of Episodic Memory*. Oxford: Clarendon Press.
84. Mercado, E., Murray, S. O., Uyeyama, R. K., Pack, A. A., & Herman, L. M. (1998). Memory for recent actions in the bottlenosed dolphin (*Tursiops truncatus*): repetition of arbitrary behaviors using an abstract rule. *Animal Learning Behavior, 26*(2), 210–18.
85. Zentall, T. R. (2008). Representing past and future events. In E. Dere, A. Easton, L. Nadel, & J. P. Huston (Eds.), *Handbook of Episodic Memory Research* (pp. 217–34). Oxford: Elsevier, 230.
86. Mercado, E., Uyeyama, R. K., Pack, A. A., & Herman, L. M. (1999). Memory for action events in the bottlenose dolphin. *Animal Cognition, 2*, 17–25.
87. For an overview see Raby, C. R., & Clayton, N. S. (2012). Episodic memory and future planning. In J. Vonk & T. Shackelford (Eds.), *Oxford Handbook of Comparative Evolutionary Psychology* (pp. 217–35). Oxford: Oxford University Press, 227–8.
88. Dally, J. M., Emery, N. J., & Clayton, N. S. (2006). Food-caching western scrub-jays keep track of who was watching when. *Science, 312*(5780), 1662–5.
89. Raby, C. R., & Clayton, N. S. (2012). Episodic memory and future planning. In J. Vonk & T. Shackelford (Eds.), *Oxford Handbook of Comparative Evolutionary Psychology* (pp. 217–35). Oxford: Oxford University Press, 227–8.
90. Suddendorf T., & Corballis M.C. (2007) The evolution of foresight: What is mental time travel and is it unique to humans? *Behavioral and Brain Sciences 30*(3), 316–17.
91. Naqshbandi, M., & Roberts, W. A. (2006). Anticipation of future events in squirrel monkeys (*Saimiri sciureus*) and rats (*Rattus norvegicus*): tests of the Bischof-Kohler hypothesis. *Journal of Comparative Psychology*, 120(4), 345–57.
92. Köhler W. (1925). *The Mentality of Apes*. New York: Harcourt, Brace & Company.
93. Osvath, M. (2009). Spontaneous planning for future stone throwing by a male chimpanzee. *Current Biology, 19*, R190–R191.
94. Osvath, M., & Karvonen, E. (2012) Spontaneous innovation for future deception in a male chimpanzee. *PLoS ONE 7*(5): e36782. doi:10.1371/journal.pone.0036782.
95. Clayton, N. S., Bussey, T. J., & Dickinson, A. (2003). Can animals recall the past and plan for the future? *Nature Reviews Neuroscience*, 4(8), 685–191.
96. Mulcahy, N. J., & Call, J. (2006). Apes save tools for future use. *Science, 312*, 1038–40.
97. Dufour, V., & Sterck, E. H. M. (2008). Chimpanzees fail to plan in an exchange task but succeed in a tool-using procedure. *Behavioural Processes, 79*(1), 19–27.
98. Osvath, M., & Osvath, H. (2008). Chimpanzee (*Pan troglodytes*) and orangutan (*Pongo abelii*) forethought: self-control and pre-experience in the face of future tool use. *Animal Cognition, 11*(4), 661–74.
99. Naqshbandi, M., & Roberts, W. A. (2006). Anticipation of future events in squirrel monkeys (*Saimiri sciureus*) and rats (*Rattus norvegicus*): tests of the Bischof-Kohler hypothesis. *Journal of Comparative Psychology, 120*(4), 345–57.

100. Raby, C. R., Alexis, D. M., Dickinson, A., & Clayton, N. S. (2007). Planning for the future by western scrub-jays. *Nature, 445*(7130), 919–21.
101. For an overview of dolphin planning see Kuczaj, S. A., II, Xitco, M. J., Jr., & Gory, J. D. (2010). Can dolphins plan their behavior? *International Journal of Comparative Psychology, 23,* 664–70. And also Kuczaj, S. A., II, Gory, J. D., & Xitco, M. J., Jr. (2009). How intelligent are dolphins? A partial answer based on their ability to plan their behavior when confronted with novel problems. *Japanese Journal of Animal Psychology, 59,* 99–115.
102. Shettleworth, S. J. (2010). Clever animals and killjoy explanations in comparative psychology. *Trends in Cognitive Sciences, 14*(11), 477–81.
103. Connor, R. C., & Krützen, M. (2003). Levels and patterns in dolphin alliance formation. In F. de Waal, & P. L. Tyack (Eds.), *Animal Social Complexity* (pp. 115–20). Cambridge, MA: Harvard University Press.
104. Visser, I. N., Smith, T. G., Bullock, I. D., Green, G. D., Carlsson, O. G., & Imberti, S. (2008). Antartic peninsula killer whales (*Orcinus orca*) hunt seals and a penguin on floating ice. *Marine Mammal Science, 24,* 225–34.
105. Smolker, R., Richards, A., Connor, R., Mann, J., & Berggren, P. (1997). Sponge carrying by dolphins (Delphinidae, *Tursiops* sp.): a foraging specialization involving tool use? *Ethology, 103,* 454–65.
106. Fertl, D., & Wilson, B. (1997). Bubble use during prey capture by a lone bottlenose dolphin (Tursiops truncatus). *Aquatic Mammals,* 23(2), 113–14.
107. Duffy-Echevarria, E. E., Connor, R. C., & Aubin, D. J. S. (2008). Observations of strand-feeding behavior by bottlenose dolphins (*Tursiops truncatus*) in Bull Creek, South Carolina. *Marine Mammal Science, 24,* 202–6.
108. Kuczaj, S. A., & Makecha, R. (2008). The role of play in the evolution and ontogeny of contextually flexible communication. In U. Griebel & K. Oller (Eds.), *Evolution of Communicative Flexibility: Complexity, Creativity, and Adaptability in Human and Animal Communication* (pp. 253–77). Cambridge, MA: MIT Press.
109. Kuczaj, S. A., II, Xitco, M. J., Jr., & Gory, J. D. (2010). Can dolphins plan their behavior? *International Journal of Comparative Psychology, 23,* 664–70.
110. Finn, J., Tregenza, T., & Norman, M. (2009). Preparing the perfect cuttlefish meal: complex prey handling by dolphins. *PLoS ONE* 4(1), e4217. doi:10.1371/journal.pone.0004217.
111. Kuczaj, S. A., II & Walker, R. T. (2012). Dolphin problem solving. In T. Zentall & E. Wasserman (Eds.), *Handbook of Comparative Cognition.* Oxford: Oxford University Press.
112. Herman, L. M. (2006). Intelligence and rational behaviour in the bottlenosed dolphin. In S. Hurley & M. Nudds (Eds.), *Rational Animals?* (pp. 439–67). Oxford: Oxford University Press, 441.
113. Kuczaj, S. A., II, Gory, J. D., & Xitco, M. J., Jr. (2009). How intelligent are dolphins? A partial answer based on their ability to plan their behavior when confronted with novel problems. *Japanese Journal of Animal Psychology, 59,* 99–115.
114. Kuczaj, S. A., II & Walker, R. T. (2012). Dolphin problem solving. In T. Zentall & E. Wasserman (Eds.), *Handbook of Comparative Cognition* (pp. 736–56). Oxford: Oxford University Press.

115. Kuczaj, S.A., II, Xitco, M. J., Jr., & Gory, J. D. (2010). Can dolphins plan their behavior? *International Journal of Comparative Psychology, 23*, 664–70. Pg. 668.
116. Foerder, P., Galloway, M., Barthel, T., Moore, D. E., & Reiss, D. (2011). Insightful problem solving in an Asian elephant. *PLoS ONE, 6*(8), 7.
117. Köhler, W. (1925). *The Mentality of Apes*. New York: Harcourt, Brace & Company.
118. Werdenich, D., & Huber, L. (2006). A case of quick problem solving in birds: string pulling in keas, *Nestor notabilis. Animal Behaviour, 71*(4), 855–63.
119. Layton, N. (2007). Animal cognition: crows spontaneously solve a metatool task. *Current Biology, 17*(20), R894–895.
120. Bird, C. D., & Emery, N. J. (2009). Insightful problem solving and creative tool modification by captive nontool-using rooks. *Proceedings of the National Academy of Sciences of the United States of America, 106*(25), 10370–5.
121. Burghardt, G. M. (2005). *The Genesis of Animal Play*. Cambridge, MA: MIT Press, xi.
122. For overview see Graham, K. L., & Burghardt, G. M. (2010). Current perspectives on the biological study of play: signs of progress. *Quarterly Review of Biology, 85*(4): 393–418.
123. Even turtles need recess: many animals—not just dogs, cats, and monkeys—need a little play time. (October 24, 2010). *ScienceDaily*. Retrieved from http://www.sciencedaily.com-/releases/2010/10/101019132045.htm.
124. Mitchell, R. W. (1990). A theory of play. In M. Bekoff & D. Jamieson (Eds.), *Interpretation and Explanation in the Study of Animal Behavior*, Vol. 1: *Interpretation, Intentionality, and Communication* (pp. 197–227). Boulder, CO: Westview Press, 197.
125. Paulos, R. D., Trone, M., Kuczaj, S. A., II (2010). Play in wild and captive cetaceans. *International Journal of Comparative Psychology, 23*, 701–22. Pg. 702.
126. Paulos, R. D., Trone, M., Kuczaj, S. A., II (2010). Play in wild and captive cetaceans. *International Journal of Comparative Psychology, 23*, 701–22. Pg. 707.
127. Herzing, D. (2011). *Dolphin Diaries: My 20 Years with Spotted Dolphins in the Bahamas*. New York: St. Martin's Press.
128. Trone, M., Kuczaj, S., & Solangi, M. (2005). Does participation in dolphin–human interaction programs affect bottlenose dolphin behaviour? *Applied Animal Behaviour Science, 93*, 363–74.
129. Slooten, E., & Dawson, S. M. (1994). Hector's dolphins. In S. H. Ridgway & R. Harrison (Eds.), *Handbook of Marine Mammals* (pp. 311–34). London: Academic Press.
130. Brown, D. H., & Norris, K. S. (1956). Observations of captive and wild cetaceans. *Journal of Mammalogy, 37*(3), 311–26.
131. Kuczaj, S. A., II & Walker, R. T. (2012). Dolphin problem solving. In T. Zentall & E. Wasserman (Eds.), *Handbook of Comparative Cognition* (pp. 736–56). Oxford: Oxford University Press.
132. Gewalt, W. (1989). Orinoco freshwater dolphins (*Inia geoffrensis*) using self-produced air bubble rings as toys. *Aquatic Mammals, 15*(2), 73–9.
133. Marten, K., Shariff, K., Psarakos, S., & White, D. J. (1996). Ring bubbles of dolphins. *Scientific American, 275*, 83–7.

134. This video can be found online by searching for "dolphin play bubble rings."
135. For an overview of all these behaviors, see Paulos, R. D., Trone, M., Kuczaj, S. A., II (2010). Play in wild and captive cetaceans. *International Journal of Comparative Psychology, 23*, 701–22.
136. Paulos, R. D., Trone, M., Kuczaj, S. A., II (2010). Play in wild and captive cetaceans. *International Journal of Comparative Psychology, 23*, 701–22. Pg. 707.
137. Delfour, F., & Aulagnier, S. (1997). Bubbleblow in beluga whales (*Delphinapterus leucas*): a play activity? *Behavioural Processes, 40*, 183–6.
138. McCowan, B., Marino, L., Vance, E., Walke, L., & Reiss, D. (2000). Bubble ring play of bottlenose dolphins (*Tursiops truncatus*): implications for cognition. *Journal of Comparative Psychology, 114*(1), 98–106.
139. Bekoff, M & Byers, J. A. (Eds.) (1998). *Animal Play: Evolutionary, Comparative, and Ecological Perspectives.* Cambridge: Cambridge University Press.
140. Burghardt, G. M. (2005). *The Genesis of Animal Play*. Cambridge, MA: MIT Press.
141. Dolphins evolve opposable thumbs. (2000, August). *The Onion*, 36(30).
142. Deecke, V. B. (2012). Tool-use in the brown bear (*Ursus arctos*). *Animal Cognition*, 15, 725–30.
143. Shumaker, R. W., Walkup, K. R., & Beck, B. B. (2011). *Animal Tool Behavior: The Use and Manufacture of Tools by Animals.* Baltimore, ML: Johns Hopkins University Press.
144. Baber, C. (2003). *Cognition and Tool Use: Forms of Engagement in Human and Animal Use of Tools.* Boca Raton, FL: CRC Press.
145. Goodall, J. (1964). Tool-using and aimed throwing in a community of free-living chimpanzees. *Nature, 201*, 1264–6.
146. Karplus, I., Fiedler, G. C., & Ramcharan, P. (1998). The intraspecific fighting behavior of the Hawaiian boxer crab, *Lybia edmondsoni*—fighting with dangerous weapons ? *Symbiosis*, 24(3), 287–301.
147. Weir, A. A. S., Chappell, J., & Kacelnik, A. (2002). Shaping of hooks in New Caledonian crows. *Science, 297*(5583), 981.
148. Shumaker, R. W., Walkup, K. R., & Beck, B. B. (2011). *Animal Tool Behavior: The Use and Manufacture of Tools by Animals.* Baltimore, ML: Johns Hopkins University Press, 5.
149. Smolker, R., Richards, A., Connor, R., Mann, J., & Berggren, P. (1997). Sponge carrying by dolphins (Delphinidae, *Tursiops* sp.): a foraging specialization involving tool use? *Ethology, 103*, 454–65.
150. Mann, J., Stanton, M. A., Patterson, E. M., Bienenstock, E. J., & Singh, L. O. (2012). Social networks reveal cultural behaviour in tool-using using dolphins. *Nature Communications*, 3, 980. doi:10.1038/ncomms1983.
151. Mann, J., Sargeant, B. L., Watson-Capps, J. J., Gibson, Q. A., Heithaus, M. R., Connor, R. C., & Patterson, E. (2008). Why do dolphins carry sponges? (R. Brooks, Ed.) *PLoS ONE*, 3(12), 7.
152. Patterson, E. M., & Mann, J. (2011). The ecological conditions that favor tool use and innovation in wild bottlenose dolphins (*Tursiops* sp.). (S. F. Brosnan, Ed.) *PLoS ONE*, 6(7), 7.

153. Krützen, M., Mann, J., Heithaus, M. R., Connor, R. C., Bejder, L., & Sherwin, W. B. (2005). Cultural transmission of tool use in bottlenose dolphins. *Proceedings of the National Academy of Sciences of the United States of America, 102*(25), 8939–43.
154. Bacher, K., Allen, S., Lindholm, A. K., Bejder, L., & Krützen, M. (2010). Genes or culture: are mitochondrial genes associated with tool use in bottlenose dolphins (*Tursiops* sp.)? *Behavior Genetics, 40*(5), 706–14.
155. Kopps, A. M., & Sherwin, W. B. (2012). Modelling the emergence and stability of a vertically transmitted cultural trait in bottlenose dolphins. *Animal Behaviour*. doi:10.1016/j.anbehav.2012.08.029.
156. See, for example, Pearson, H. C., & Shelton, D. E. (2010). A large-brained social animal. In B. Würsig & M. Würsig (Eds.), *The Dusky Dolphin: Master Acrobat off Different Shores* (pp. 333–53). San Diego, CA: Elsevier.
157. Thouless, C. R., Fanshawe, J. H., & Bertram, B. C. R. (1989). Egyptian vultures *Neophron percnopterus* and ostrich *Struthio camelus* eggs: the origins of stone-throwing behavior. *Ibis, 131*(1), 9–15.
158. Visalberghi, E., Fragaszy, D., Ottoni, E., Izar, P., De Oliveira, M. G., & Andrade, F. R. D. (2007). Characteristics of hammer stones and anvils used by wild bearded capuchin monkeys (*Cebus libidinosus*) to crack open palm nuts. *American Journal of Physical Anthropology, 132*(3), 426–44.
159. Cetacean Rights: Conference on Fostering Moral and Legal Change, Helsinki Collegium for Advanced Studies, University of Helsinki, Finland, May 21, 2010.
160. Laland, K. N., &. and Galef, B. G. (Eds.) (2009). *The Question of Animal Culture*. Cambridge, MA: Harvard University Press.
161. Sargeant, B. L., & Mann, J. (2009). Social learning to culture: intrapopulation variation in bottlenose dolphins. In K. N. Laland & B. G. Galef, *The Question of Animal Culture* (pp. 152–73). Cambridge, MA: Harvard University Press.
162. Tomasello, M. (2009). The question of chimpanzee culture, plus postscript (Chimpanzee culture, 2009). In K. N. Laland & B. G. Galef (Eds.), *The Question of Animal Culture* (pp. 198–221). Cambridge, MA: Harvard University Press.
163. Lumsden, C. J., & Wilson, E. O. (1981). *Genes, Mind, and Culture: The Coevolutionary Process*. Cambridge, MA: Harvard University Press.
164. Rendell, L., & Whitehead, H. (2001). Culture in whales and dolphins. *Behavioral and Brain Sciences, 24*(2), 309–24; discussion 324–82.
165. This definition was taken from: Boyd, R., & Richerson, P. J. (1996). Why culture is common but cultural evolution is rare. *Proceedings of the British Academy, 88*, 77–93.
166. Mann, J. (2001). Cetacean culture: definitions and evidence. *Behavioral and Brain Sciences, 24*(2), 343.
167. See Whitehead, H. (2009). How might we study culture: a perspective from the ocean. In K. N. Laland & B. G. Galef (Eds.), *The Question of Animal Culture* (pp. 125–51). Cambridge, MA: Harvard University Press, 149.
168. Noad, M. J., Cato, D. H., Bryden, M. M., Jenner, M. N., & Jenner, K. C. (2000). Cultural revolution in whale songs. *Nature, 408*(6812), 537.
169. Whitehead, H. (2009). How we might study culture: a perspective from the ocean. In K. Laland & B. G. Galet (Eds.), *The Question of Animal Culture* (pp. 125–51). Cambridge, MA: Harvard University Press, 126.

170. Mann, J. (2001). Cetacean culture: definitions and evidence. *Behavioral and Brain Sciences, 24*(2), 343.
171. Janik, V. M. (2001). Is cetacean social learning unique? *Behavioral and Brain Sciences, 24*(2), 337–8.
172. Donaldson, R., Finn, H., Bejder, L., Lusseau, D., & Calver, M. (2012). The social side of human–wildlife interaction: wildlife can learn harmful behaviours from each other. *Animal Conservation, 15*(5), 427–35.
173. Janik, V. M. (2001). Is cetacean social learning unique? *Behavioral and Brain Sciences, 24*(2), 337–8.
174. Kuczaj, S. (2001). Cetacean culture: slippery when wet. *Behavioral and Brain Sciences,* 24, 340–1.
175. Sargeant, B. L., & Mann, J. (2009). From social learning to culture: intrapopulation variation in bottlenose dolphins. In K. N. Laland and B. G. Galef. (Eds.), *The Question of Animal Culture* (pp. 152–73). Cambridge, MA: Harvard University Press.
176. Nowacek, D. P. (2002). Sequential foraging behaviour of bottlenose dolphins, *Tursiops truncates,* in Sarasota Bay, Florida. *Behaviour* 139(9), 1125–45.
177. Connor, R. C., Heithaus, M., Berggren, P., & Miksis, J. L. (2000). "Kerplunking": Surface fluke splashes during shallow-water bottom foraging by bottlenose dolphins. *Marine Mammal Science, 16,* 646–53.
178. Sargeant, B. L., Mann, J., Berggren, P., & Krützen, M. (2005). Specialization and development of beach hunting, a rare foraging behavior, by wild bottlenose dolphins (*Tursiops* sp.). *Canadian Journal of Zoology, 83*(11), 1400–10.
179. Sargeant, B. L., & Mann, J. (2009). Developmental evidence for foraging traditions in wild bottlenose dolphins. *Animal Behaviour, 78*(3), 715–21.
180. Mann, J. (2001). Cetacean culture: definitions and evidence. *Behavioral and Brain Sciences, 24*(2), 343.
181. Galef, B.G., Jr. (2001). Where's the beef? Evidence of culture, imitation and teaching in cetaceans. *Behavioral and Brain Sciences, 24*(2), 335.
182. Premack, D., & Hauser, M. D. (2001). A whale of a tale: calling it culture doesn't help. *Behavioral and Brain Sciences,* 24(2), 350–1.
183. Mitchell, R. W. (2001). On not drawing the line about culture: inconsistencies in interpretation of nonhuman cultures. *Behavioral and Brain Sciences,* 24(2), 348.
184. Bender, C., Herzing, D., & Bjorklund, D. (2009). Evidence of teaching in Atlantic spotted dolphins (*Stenella frontalis*) by mother dolphins foraging in the presence of their calves. *Animal Cognition, 12,* 43–53.
185. Pearson, H. C., & Shelton, D. E. (2010). A large-brained social animal. In B. Würsig, & M. Würsig (Eds.), *The Dusky Dolphin: Master Acrobat off Different Shores* (pp. 333–53). San Diego, CA: Elsevier.
186. For overview see Sargeant, B. L., & Mann, J. (2009). Developmental evidence for foraging traditions in wild bottlenose dolphins. *Animal Behaviour, 78*(3), 715–21.
187. Whitehead, H. (2011). The culture of whales and dolphins. In P. Brakes & M. Simmonds (Eds.), *Whales and Dolphins: Cognition, Culture, Conservation and Human Perceptions* (pp. 149–69). London: Earthscan.

188. Laland, K. N., & Janik, V. M. (2006). The animal cultures debate. *Trends in Ecology & Evolution, 21*(10), 542–7.
189. See, for example, Mann, J., Stanton, M. A., Patterson, E. M., Bienenstock, E. J., & Singh, L. O. (2012). Social networks reveal cultural behaviour in tool-using using dolphins. *Nature Communications, 3*, 980. doi:10.1038/ncomms1983.
190. Sargeant, B. L., & Mann, J. (2009). Social learning to culture: intrapopulation variation in bottlenose dolphins. In K. N. Laland & B. G. Galef (Eds.), *The Question of Animal Culture* (pp. 152–73). Cambridge, MA: Harvard University Press.
191. Mann, J., Stanton, M. A., Patterson, E. M., Bienenstock, E. J., & Singh, L. O. (2012). Social networks reveal cultural behaviour in tool-using using dolphins. *Nature Communications, 3*, 980. doi:10.1038/ncomms1983.
192. Kuczaj, S. (2001). Cetacean culture: slippery when wet. *Behavioral and Brain Sciences*, 24, 340–1. Pg. 341.
193. Franks, N. R., & Richardson, T. (2006). Teaching in tandem-running ants. *Nature, 439*(7073), 153.
194. The authors of the ant tandem running article used this definition of "teaching": "An individual is a teacher if it modifies its behaviour in the presence of a naive observer, at some initial cost to itself, in order to set an example so that the other individual can learn more quickly." From Caro, T. M., & Hauser, M. D. (1992). Is there teaching in nonhuman animals? *Quarterly Review of Biology, 61*, 151–74.
195. Thornton, A., & Malapert, A. (2009). Experimental evidence for social transmission of food acquisition techniques in wild meerkats. *Animal Behaviour, 78*, 255–64.
196. Thornton, A., & McAuliffe, K. (2006). Teaching in wild meerkats. *Science, 313*, 227–9.
197. Wilkinson, A., Kuenstner, K., Mueller, J., & Huber, L. (2010). Social learning in a non-social reptile (*Geochelone carbonaria*). *Biology Letters, 6*(5), 614–16. doi:0.1098/rsbl.2010.0092.
198. Brown, C., & Laland, K. N. (2003). Social learning in fishes: a review. *Fish and Fisheries*, 4(3), 280–8.
199. Manassa, R. P., & McCormick, M. I. (2012). Social learning and acquired recognition of a predator by a marine fish. *Animal Cognition*, 15(4), 559–65.
200. Warner, R. R. (1988). Traditionality of mating-site preferences in a coral reef fish. *Nature, 335*, 719–72.
201. Guttridge, T. L., van Dijk, S., Stamhuis, E. J., Krause, J., Gruber, S. H., & Brown, C. (2012). Social learning in juvenile lemon sharks, *Negaprion brevirostris. Animal Cognition*. doi:10.1007/s10071-012-0550-6.
202. Whiten, A., Goodall, J., McGrew, W. C., Nishida, T., Reynolds, V., Sugiyama, Y., Tutin, C. E., et al. (1999). Cultures in chimpanzees. *Nature*, 399(6737), 682–5.
203. Gruber, T., Muller, M. N, Strimling, P., Wrangham, R., & Zuberbuhler, K. (2009). Wild chimpanzees rely on cultural knowledge to solve an experimental honey acquisition task. *Current Biology, 19*, 1806–10.
204. Nakamura, M., & Uehara, S. (2004). Proximate factors of two types of grooming-hand-clasp in Mahale chimpanzees: implication for chimpanzee social custom. *Current Anthropology, 45*(1), 108–14.

205. Whiten, A., & Van Schaik, C. P. (2007). The evolution of animal "cultures" and social intelligence. *Philosophical Transactions of the Royal Society of London—Series B: Biological Sciences, 362*(1480), 603–20.
206. Slater, P. (2003). Fifty years of bird song research: a case study in animal behaviour. *Animal Behaviour, 65*(4), 633–9.
207. Laland, K. N., & Janik, V. M. (2006). The animal cultures debate. *Trends in Ecology & Evolution, 21*(10), 542–7.
208. Whitehead, H. (2011). The culture of whales and dolphins. In P. Brakes & M. Simmonds (Eds.), *Whales and Dolphins: Cognition, Culture, Conservation and Human Perceptions* (pp. 149–69). London: Earthscan.

Chapter 5

1. Fitch, W. T. (2010). *The Evolution of Language*. Cambridge: Cambridge University Press, 148.
2. Pinker, S. (1994). *The Language Instinct*. New York: William Morrow, 334
3. Anderson, S. (2006). *Doctor Dolittle's Delusion: Animals and the Uniqueness of Human Language*. New Haven, CT: Yale University Press, 2.
4. Deacon, T. (1997). *The Symbolic Species: The Co-Evolution of Language and the Brain*. London: Penguin Books, 25.
5. Bickerton, D. (2009). *Adam's Tongue: How Humans Made Language, How Language Made Humans*. New York: Hill and Wang, 4.
6. Corballis, M. (2002). *From Hand to Mouth: The Origins of Language*. Princeton, NJ: Princeton University Press, 10.
7. Chomsky, N. (2006). *Language and the Mind*. New York: Cambridge University Press, 59.
8. Hobaiter, C., & Byrne, R. (2011). The gestural repertoire of the wild chimpanzee. *Animal Cognition, 14*, 745–67. Pg. 745.
9. The opening line to Jane Austen's *Pride and Prejudice*.
10. With Con Slobodchikoff being a notable exception, and representing what should perhaps be considered the minority opinion in the current scientific community. See Slobodchikoff, C. (2012). *Chasing Doctor Dolittle: Learning the Language of Animals*. New York: St. Martin's Press.
11. "Within the next decade or two the human species will establish communication with another species: nonhuman, alien, possibly extraterrestrial, more probably marine." Lilly, J. (1962). *Man and Dolphin*. London: Victor Gollancz, 15.
12. See discussion of this time period in Chapter 2 of Reynolds, J. E., III, Wells, R. S., and Eide, S. D. (2000). *Biology and Conservation of the Bottlenose Dolphin*. Gainsville, FL: University Press of Florida, with special reference to the following article: Caldwell, D. K. & Caldwell, M. C. (1972). Dolphins communicate—but they don't talk. *Naval Research Reviews* (June–July), 23–7.
13. Lilly, J. C. (1962). *Man and Dolphin*. London: Victor Gollancz.
14. Lilly, J. C. (1967). *The Mind of the Dolphin: A Nonhuman Intelligence*. New York: Doubleday.

15. Lilly, J. C. (1978). *Communication Between Man and Dolphin: The Possibilities of Talking with Other Species.* New York: Crown Publishers.
16. Munkittrick, K. (February 18, 2011). Learning the alien language of dolphins. Retrieved from http://blogs.discovermagazine.com/sciencenotfiction/2011/02/18/learning-the-alien-language-of-dolphins/.
17. Scott, K. (September 8, 2011). Dolphins may "talk" like humans. *Wired Science* Retrieved from http://www.wired.com/wiredscience/2011/09/dolphin-language/.
18. Australian researcher partly decodes dolphin language. (December 20, 2007). *Fox News* Retrieved from http://www.foxnews.com/story/0,2933,317471,00.html.
19. Viegas, J. (February 28, 2012). Dolphins greet each other at sea. *Discovery News.* Retrieved from http://news.discovery.com/animals/dolphins-greet-each-other-120228.html.
20. Sirius Institute FAQ (April 24, 2005). Retrieved from http://www.planetpuna.com/faq.htm.
21. Global Heart, Inc. (2011). The discovery of dolphin language [Press Release]. Retrieved from http://www.speakdolphin.com/ResearchItems.cfm?ID=20.
22. Sagan, C., & Agel, J. (1973). *Cosmic Connection: An Extraterrestrial Perspective.* Garden City, NY: Anchor Press, 177.
23. Sagan, C., & Agel, J. (1973). *Cosmic Connection: An Extraterrestrial Perspective.* Garden City, NY: Anchor Press, 179.
24. Lieberman, P. (1989). Some biological constraints on universal grammar and learnability. In M. Rice & R. Schiefelbusch (Eds.), *The Teachability of Language* (pp. 199–235). Baltimore, ML: Paul. H. Brookes, 222.
25. The following quote is often attributed to Carl Sagan, and was published in Gaither, C. C., & Cavazos-Gaither, A. E. (2008). *Gaither's Dictionary of Scientific Quotations.* New York: Springer. "It is of interest to note that while some dolphins are reported to have learned English—up to fifty words used in correct context—no human being has been reported to have learned dolphinese." The claimed source (according to *Gaither's Dictionary*) is from Sagan's essay "The Burden of Skepticism" from *Skeptical Enquirer*, 12, Fall 1987, 38–46, Pg. 46. However, this quote does not actually appear in this article from *Skeptical Inquirer*, and cannot therefore be reliably attributed to Sagan. I could find no evidence for the true source of this quote.
26. See, for example, Kuczaj, S. A., & Kirkpatrick, V. M. (1993). Similarities and differences in human and animal language research: toward a comparative psychology of language. In H. L. Roitblat, L. M. Herman, & P. E. Nachtigall (Eds.), *Language and Communication: Comparative Perspectives* (pp. 45–63). Hillsdale, NJ: Lawrence Erlbaum Associates, 46.
27. After I wrote this section using the traffic lights analogy, I read *Doctor Doolittle's Delusion* by Yale professor Stephen R. Anderson and found out (to my surprise) that he had used the same exact analogy. Thus, credit for the traffic light analogy remains with Professor Anderson: Anderson, S. (2006). *Doctor Dolittle's Delusion: Animals and the Uniqueness of Human Language.* New Haven, CT: Yale University Press.

28. Bickerton, D. (2009). *Adam's Tongue: How Humans Made Language, How Language Made Humans*. New York: Hill and Wang.
29. Deacon, T. (1997). *The Symbolic Species: The Co-Evolution of Language and the Brain*. London: Penguin Books.
30. Corballis, M. C. (2011). *The Recursive Mind: The Origins of Human Language, Thought and Civilization*. Princeton, NJ: Princeton University Press.
31. Bickerton, D. (2009). *Adam's Tongue: How Humans Made Language, How Language Made Humans*. New York: Hill and Wang.
32. Tomasello, M. (2008). *Origins of Human Communication*. Cambridge, MA: MIT Press.
33. Lieberman, P. (2006). Toward an evolutionary biology of language. *Science, 314*(5801), 926–7.
34. Dunbar, R. (1996). *Grooming, Gossip, and the Evolution of Language*. Cambridge, MA: Harvard University Press.
35. Pinker, S. (1994). *The Language Instinct*. New York: William Morrow.
36. Fitch, W. T. (2004). Kin selection and "mother tongues": a neglected component in language evolution. In D. Kimbrough Oller & U. Griebel (Eds.), *Evolution of Communication Systems: A Comparative Approach* (pp. 275–96). Cambridge, MA: MIT Press.
37. Chomsky, N. (2007). Of minds and language. *Biolinguistics 1*, 9–27.
38. Hockett, C. F., & Altmann, S. (1968). A note on design features. In T. A. Sebeok (Ed.), *Animal Communication: Techniques of Study and Results of Research* (pp. 61–72). Bloomington: Indiana University Press.
39. For example, this highly influence article: Hauser, M. D, Chomsky, N., & Fitch, W. T. (2002). The faculty of language: what is it, who has it, and how did it evolve? *Science, 298* (5598), 1569–79.
40. Darwin, C. (1871). *The Descent of Man, and Selection in Relation to Sex*. London: John Murray, 120.
41. Penn, D. C., Holyoak, K. J., & Povinelli, D. J. (2008). Darwin's mistake: explaining the discontinuity between human and nonhuman minds. *Behavioral and Brain Sciences, 31*(2), 109–30.
42. Penn, D. C., Holyoak, K. J., & Povinelli, D. J. (2008). Darwin's mistake: explaining the discontinuity between human and nonhuman minds. *Behavioral and Brain Sciences, 31*(2), 109–78. Pg. 154.
43. Herman, L. M., Uyeyama, R. K., & Pack, A. A. (2008). Bottlenose dolphins understand relationships between concepts. *Behavioral and Brain Sciences, 31*(2), 139–40.
44. For a discussion see Roitblat, H. L., Harley, H. E., & Helweg, D. A. (1993). Cognitive processing in artificial language research. In H. L. Roitblat, L. M. Herman, & P. E. Nachtigall (Eds.), *Language and Communication: Comparative Perspectives* (pp. 1–23). Hillsdale, NJ: Lawrence Erlbaum Associates.
45. Bastian, J. (1967). The transmission of arbitrary environmental information between bottlenosed dolphins. In R. G. Busnel (Ed.), *Animal Sonar Systems* (pp. 803–73). New York: Plenum Press.
46. Bastian, J., Wall, C., & Anderson, C. L. (1968). Further investigation of the transmission of arbitrary environmental information between bottle-nose dolphins. Naval Undersea Warfare Center. Report no. TP 109, 38.

47. Herman, L. M., & Tavolga, W. N. (1980). The communication systems of cetaceans. In L. M. Herman (Ed.), *Cetacean Behavior: Mechanisms and Functions* (pp. 149–209). New York: Wiley Interscience.
48. Evans, W. E., & Bastian, J. (1969). Marine mammal communication; social and ecological factors. In H. T. Anderson (Ed.), *The Biology of Marine Mammals* (pp. 425–76). New York: Academic Press.
49. Dudok van Heel, W. H. (1974). *Extraordinaires Dauphins.* Paris: Rossel.
50. Herman, L. M. (2009). Language learning and cognitive skills. In W. F. Perrin, B. Würsig, & H. C. M. Thewissen (Eds.), *Encyclopedia of Marine Mammals,* 2nd edn (pp. 657–63). New York: Academic Press.
51. The following sentence: "In a noted experiment Dr. Javis Bastian found that dolphins could communicate about abstract ideas," appears in What makes dolphins so smart? (n.d.) Retrieved from http://animal.discovery.com/features/dolphins/article/article.html. More than likely the Discovery Channel sourced their information from the following book which uses almost identical wording on page 108 when discussing this topic: Malone, J. (2001). *Unsolved Mysteries of Science: A Mind-Expanding Journey through a Universe of Big Bangs, Particle Waves, and Other Perplexing Concepts.* New York: John Wiley & Sons.
52. Consider this quote: "Obviously Doris must have been telling Buzz that the light was either flashing or continuous for him to know which was the correct lever to press," from Cochrane, A., & Callen, K. *Dolphins and Their Power to Heal.* Rochester, VT: Healing Arts Press, 88.
53. These articles are "the most important ever written in this end of the Century and could be compared to a Copernical revolution," according to http://www.dauphinlibre.be/langintro.htm.
54. Zanin, A. V., Markov, V. I., & Sidorova, I. E. (1990). The ability of bottlenose dolphins, *Tursiops truncatus,* to report arbitrary information. In J. A. Thomas & R. A. Kastelein (Eds.), *Sensory Abilities of Cetaceans—Laboratory and Field Evidence* (pp. 685–97). NATO ASI Series, Series A: Life Sciences, Vol. 196. New York: Plenum Press.
55. Markov, V. I., & Ostrovskaya, V. M. (1990). Organization of communication system in *Tursiops truncatus* Montagu. In J. A. Thomas & R. A. Kastelein (Eds.), *Sensory Abilities of Cetaceans—Laboratory and Field Evidence* (pp. 599–602). NATO ASI Series, Series A: Life Sciences, Vol. 196. New York: Plenum Press.
56. Ivanov, P. M. (2009). Study of dolphin communicational behavior: procedure, motor and acoustic parameters. *Journal of Evolutionary Biochemistry and Physiology,* 45(6), 696–705. Original Russian text published in: (2009) *Zhurnal Evolyutsionnoi Biokhimii i Fiziologii,* 45(6), 575–82.
57. Cooper, K. (August 29, 2012). Dolphins, aliens, and the search for intelligent life. Retrieved from http://www.astrobio.net/exclusive/4182/dolphins-aliens-and-the-search-for-intelligent-life-.
58. Bshary, R., Hohner, A., Ait-el-Djoudi, K., & Fricke, H. (2006). Interspecific communicative and coordinated hunting between groupers and giant moray eels in the Red Sea. *PLoS Biology,* 4(12), e431. doi:10.1371/journal.pbio.0040431.

59. Couzin, I. D. (2006). Behavioral ecology: social organization in fission–fusion societies. *Current Biology, 16*(5), R169–R171.
60. Slobodchikoff, C. N., Perla, B. S., & Verdolin, J. L. (2009). *Prairie Dogs: Communication and Community in an Animal Society*. Cambridge, MA: Harvard University Press.
61. Dreher, J. J. (1966). Cetacean communication: small-group experiment. In K. Norris (Ed.), *Whales, Dolphins and Porpoises* (pp. 529–43). Berkeley and Los Angeles: University of California Press.
62. Dreher, J. J. (1961). Linguistic considerations of porpoise sounds. *Journal of the Acoustical Society of America, 33*, 1799–800.
63. Dreher, J. J., & Evans, W. E. (1964). Cetacean communication. In W. N. Tavolga (Ed.), *Marine Bio-Acoustics* (pp. 473–393). Oxford: Pergamon.
64. Lang, T. G., & Smith, H. A. P. (1965). Communication between dolphins in separate tanks by way of an acoustic link. *Science, 150*, 1839–43.
65. Lilly, J. C., & Miller, A. M. (1961). Vocal exchanges between dolphins. *Science*, 134, 1873–76.
66. Dreher, J. J. (1966). Cetacean communication: small-group experiment. In K. Norris (Ed.), *Whales, Dolphins and Porpoises* (pp. 529–43). Berkeley and Los Angeles: University of California Press, 542.
67. Herman, L. M. and Tavolga, W. N. (1980). The communication systems of cetaceans. In L. M. Herman (Ed.), *Cetacean Behavior: Mechanisms and Functions* (pp. 149–209). New York: Wiley Interscience.
68. For example, Hawkins, E., & Gartside, D. (2009). Patterns of whistles emitted by wild Indo-Pacific bottlenose dolphins (*Tursiops aduncus*) during a provisioning program. *Aquatic Mammals, 35*(2), 171–86.
69. Harley, H. E. (2008). Whistle discrimination and categorization by the Atlantic bottlenose dolphin (*Tursiops truncatus*): a review of the signature whistle framework and a perceptual test. *Behavioural Processes, 77*, 243–68.
70. Hawkins, E., & Gartside, D. (2010). Whistle emissions of Indo-Pacific bottlenose dolphins (*Tursiops aduncus*) differ with group composition and surface behaviours. *Journal of the Acoustical Society of America, 127*(4), 2652–63.
71. Hernandez, E. N., Solangi, M., & Kuczaj, S. A. (2010). Time and frequency parameters of bottlenose dolphin whistles as predictors of surface behavior in the Mississippi Sound. *Journal of the Acoustical Society of America, 127*(5), 3232–8.
72. Azevedo, A. F., Flach, L., Bisi, T. L., Andrade, L. G., Dorneles, P. R., & Lailson-Brito, J. (2010). Whistles emitted by Atlantic spotted dolphins (*Stenella frontalis*) in Southeastern Brazil. *Journal of the Acoustical Society of America, 127*(4), 2646–51.
73. Janik, V. M. (2009). Acoustic communication in delphinids. In M. Naguib & V. M. Janik (Eds.), *Advances in the Study of Behavior*, Vol. 40 (pp. 123–57). Oxford: Elsevier. Pg. 129.
74. Janik, V. M. (2013). Cognitive skills in bottlenose dolphin communication. *Trends in Cognitive Science* (In Press).
75. Kako, E. (1999). Elements of syntax in the systems of three language-trained animals. *Animal Learning and Behavior, 27*, 1–14.
76. Hillix, W. A., & Rumbaugh, D. (2004). *Animal Bodies Human Minds. Ape, Dolphin, and Parrot Language Skills*. New York: Plenum Press.

77. Ramos, D., & Ades, C. (2012). Two-item sentence comprehension by a dog (*Canis familiaris*). *PLoS ONE 7*(2), e29689. doi:10.1371/journal.pone.0029689.
78. Akmajian, A., Demers, R. A., Farmer, A. K., & Harnish, R. M. (2010). *Linguistics: An Introduction to Language and Communication*. Cambridge, MA: MIT Press, 211.
79. Hauser, M. D, Chomsky, N., & Fitch, W. T. (2002). The faculty of language: what is it, who has it, and how did it evolve? *Science, 298*(5598), 1569–79.
80. Herman, L. M., Richards, D. G., & Wolz, J. P. (1984). Comprehension of sentences by bottlenosed dolphins. *Cognition, 16,* 129–219.
81. Herman, L. M. (1987). Receptive competences of language-trained animals. In J. S. Rosenblatt, C. Beer, M. C. Busnel, & P. J. B. Slater (Eds.), *Advances in the Study of Behavior*, Vol. 17 (pp. 1–60). Petaluma, CA: Academic Press.
82. Penn, D. C., Holyoak K. J., & Povinelli, D. J. (2008). Darwin's mistake: explaining the discontinuity between human and nonhuman minds. *Behavioral and Brain Sciences, 31*(2), 109–30.
83. Fedor, A., Ittzés, P., & Szathmáry, E. (2010). Parsing recursive sentences with a connectionist model including a neural stack and synaptic gating. *Journal of Theoretical Biology, 271,* 100–5.
84. Luuk, E., & Luuk, H. (2011). The redundancy of recursion and infinity for natural language. *Cognitive Processing, 12*(1), 1–11.
85. Corballis, M. C. (2007). Recursion, language, and starlings. *Cognitive Science, 31,* 697–704.
86. Suzuki, R., Buck, J. R., & Tyack, P. L. (2006). Information entropy of humpback whale songs. *Journal of the Acoustical Society of America, 119*(3), 1849–66.
87. Abe, K., & Watanabe, D. (2011). Songbirds possess the spontaneous ability to discriminate syntactic rules. *Nature Neuroscience, 14,* 1067–74.
88. Smith, J., Goldizen, A. W., Dunlap, R. A., & Noad, M. J. (2008). Songs of male humpback whales, *Megaptera novaeangliae*, are involved in intersexual interactions. *Animal Behaviour 76,* 467–77.
89. Handel, S., Todd, S. K., & Zoidis, A. M. (2012). Hierarchical and rhythmic organization in the songs of humpback whales (*Megaptera novaeangliae*). *Bioacoustics, 21*(2), 141–56.
90. Zechmeister, E. B., Chronis, A. M., Cull, W. L., D'Anna, C. A., & Healy, N. A. (1995). Growth of a functionally important lexicon. *Journal of Reading Behavior, 27*(2), 201–12.
91. Diller, K. C. (1978). *The Language Teaching Controversy*. Rowley, MA: Newbury House.
92. Cheney, D. L., & Seyfarth, R. M. (1990). *How Monkeys See the World: Inside the Mind of Another Species*. Chicago, IL: University of Chicago Press.
93. For an overview see Hauser, M.D. (1996). *The Evolution of Communication*. Cambridge, MA. MIT Press.
94. Caro, T. M. (2005). Antipredator defenses in birds and mammals. *The Auk, 123,* 612.
95. Lyn, H. (2007). Mental representation of symbols as revealed by vocabulary errors in two bonobos (*Pan paniscus*). *Animal Cognition, 10*(4), 461–75.
96. Kaminski, J., Call, J., & Fischer, J. (2004). Word learning in a domestic dog: evidence for "fast mapping." *Science, 304*(5677), 1682–3.

97. Carey, B. (September 10, 2007). Alex, a parrot who had a way with words, dies. New York Times.
98. Herman, L. (n.d.) Retrieved from http://www.dolphin-institute.org/resource_guide/animal_language.htm.
99. Pilley, J. W., & Reid, A. K. (2011). Border Collie comprehends object names as verbal referents. Behavioural Processes, 86(2), 184–95.
100. Herman, L. M., & Forestell, P. H. (1985). Reporting presence or absence of named objects by a language-trained dolphin. *Neuroscience and Biobehavioral Reviews, 9*, 667–91.
101. Tyack, P. L. (1993). Animal language research needs a broader comparative and evolutionary framework. In H. L. Roitblat, L. M. Herman, & P. E. Nachtigall (Eds.), *Language and Communication: Comparative Perspectives* (pp. 115–52). Hillsdale, NJ: Lawrence Erlbaum Associates.
102. Janik, V. M., Sayigh, L. S., & Wells, R. S. (2006). Signature whistle shape conveys identity information to bottlenose dolphins. *Proceedings of the National Academy of Sciences of the United States of America, 103*(21), 8293–7.
103. Jaakola, K. (2012). Cetacean cognitive specializations. In J. Vonk & T. Shackelford (Eds.), *Oxford Handbook of Comparative Evolutionary Psychology* (pp. 144–65). New York: Oxford University Press.
104. Jarvis, E. D. (2004). Learned birdsong and the neurobiology of human language. *Annals of the New York Academy of Sciences, 1016*(1), 749–77.
105. Ralls, K., Fiorelli, P., & Gish, S. (1985). Vocalizations and vocal mimicry in captive harbor seals, *Phoca vitulina*. *Canadian Journal of Zoology, 63*, 1050–6.
106. Richards, D. G., Wolz, J. P., & Herman, L. M. (1984). Vocal mimicry of computer generated sounds and vocal labeling of objects by a bottlenosed dolphin, *Tursiops truncatus*. *Journal of Comparative Psychology, 98*, 10–28.
107. Reiss, D., & McCowan, B. (1993). Spontaneous vocal mimicry and production by bottlenose dolphins (*Tursiops truncatus*): evidence for vocal learning. *Journal of Comparative Psychology, 107*(3), 301–12.
108. Miksis, J. L., Tyack, P. L., & Buck, J. R. (2002). Captive dolphins, *Tursiops truncatus*, develop signature whistles that match acoustic features of human-made model sounds. *Journal of the Acoustical Society of America, 112*(2), 728–39.
109. Tyack, P. L. (1997). Development and social functions of signature whistles in bottlenose dolphins *Tursiops truncatus*. *Bioacoustics, 8*(1–2), 21–46.
110. Sayigh, L. S., Tyack, P. L., Wells, R. S., & Scott, M. D. (1990). Signature whistles of free-ranging bottlenose dolphins, *Tursiops truncatus*: stability and mother-offspring comparisons. *Behavioral Ecology and Sociobiology, 26*, 247–60.
111. Smolker, R., & Pepper, J. W. (1999). Whistle convergence among allied male bottlenose dolphins (Delphinidae, *Tursiops* sp.). *Ethology, 105*(7), 595–617.
112. Reiss, D., McCowan, B., & Marino, L. (1997). Communicative and other cognitive characteristics of bottlenose dolphins. *Trends in Cognitive Sciences, 1*(4), 140–5.
113. Ford, J. K. B. (1991). Vocal traditions among resident killer whales (*Orcinus orca*) in coastal waters of British Columbia. *Canadian Journal of Zoology/Revue Canadienne de Zoologie, 69*(6), 1454–83.

114. Ford, J. K. B. (1991). Vocal traditions among resident killer whales (*Orcinus orca*) in coastal waters of British Columbia. *Canadian Journal of Zoology/Revue Canadienne de Zoologie, 69*(6), 1454–83.
115. Filatova, O. A., Burdin, A. M., & Hoyt, E. (2011). Horizontal transmission of vocal traditions in killer whale (*Orcinus orca*) dialects. *Biology Bulletin, 37*(9), 965–71.
116. Seyfarth, R. M., & Cheney, D. L. (2006). Meaning and emotion in animal vocalizations. *Annals of the New York Academy of Sciences, 1000*(1), 32–55.
117. Griffin, D. R. (2001). *Animal Minds: Beyond Cognition to Consciousness*. Chicago, IL: University of Chicago Press, 165.
118. Hagenaars, M. A., & Van Minnen, A. (2005). The effect of fear on paralinguistic aspects of speech in patients with panic disorder with agoraphobia. *Journal of Anxiety Disorders, 19*(5), 521–37.
119. Esch, H. C., Sayigh, L. S., & Wells, R. S. (2009). Quantifying parameters of bottlenose dolphin signature whistles. *Marine Mammal Science, 25*, 976–86.
120. Hawkins, E. R., & Gartside, D. F. (2009). Patterns of whistles emitted by wild Indo-Pacific bottlenose dolphins (*Tursiops aduncus*) during a provisioning program. *Aquatic Mammals, 35*(2), 171–86.
121. Reiss, D., & McCowan, B. (1993). Spontaneous vocal mimicry and production by bottlenose dolphins (*Tursiops truncatus*): evidence for vocal learning. *Journal of Comparative Psychology, 107*(3), 301–12.
122. Reiss, D., & McCowan, B. (1993). Spontaneous vocal mimicry and production by bottlenose dolphins (*Tursiops truncatus*): evidence for vocal learning. *Journal of Comparative Psychology, 107*(3), 301–12.
123. Marler, P., & Evans, C. (1996). Bird calls: just emotional displays or something more? *Ibis, 138*, 26–33.
124. Radford, A. N., & Ridley, A. R. (2006). Recruitment calling: a novel form of extended parental care in an altricial species. *Current Biology, 16*(17), 1700–4.
125. Doutrelant, C., McGregor, P. K., & Oliveira, R. F. (2001). The effect of an audience on intrasexual communication in male Siamese fighting fish, *Betta splendens*. *Behavioral Ecology, 12*(3), 283–6.
126. Crockford, C., Wittig, R. M., Mundry, R., & Zuberbühler, K. (2011). Wild chimpanzees inform ignorant group members of danger. *Current Biology, 22*(2), 142–46. Pg. 142.
127. Information on the bee waggle dance was sourced from Anderson, S. (2006). *Doctor Dolittle's Delusion: Animals and the Uniqueness of Human Language*. New Haven, CT: Yale University Press.
128. Roitblat, H. L., Harley, H. E., & Helweg, D. A. (1993). Cognitive processing in artificial language research. In H. L. Roitblat, L. M. Herman, & P. E. Nachtigall (Eds.), *Language and Communication: Comparative Perspectives* (pp. 1–23). Hillsdale, NJ: Lawrence Erlbaum Associates, 2.
129. Kuczaj, S. A., & Kirkpatrick, V. M. (1993). Similarities and differences in human and animal language research: toward a comparative psychology of language. In H. L. Roitblat, L. M. Herman, & P. E. Nachtigall (Eds.), *Language and Communication: Comparative Perspectives* (pp. 45–63). Hillsdale, NJ: Lawrence Erlbaum Associates, 48.

130. Roitblat, H. L., Harley, H. E., & Helweg, D. A. (1993). Cognitive processing in artificial language research. In H. L. Roitblat, L. M. Herman, & P. E. Nachtigall (Eds.), *Language and Communication: Comparative Perspectives* (pp. 1–23). Hillsdale, NJ: Lawrence Erlbaum Associates, 6.
131. Sirius Institute FAQ (April 24, 2005). Retrieved from http://www.planetpuna.com/faq.htm.
132. Fulton, J. T. (September 9, 2009). Appendix U: Dolphin Language & Speech. Retrieved from http://neuronresearch.net/dolphin/pdf/Dolphin_language.pdf.
133. White, T. (2007). *In Defense of Dolphins: The New Moral Frontier*. Malden, MA: Blackwell Publishing, 115.
134. Herzing, D. L., & White, T. (1999). Dolphins and the question of personhood. *Etica Animali, 9*(98), 64–84. Pg. 75.
135. Lori Marino quoted in White, T. (2007). *In Defense of Dolphins: The New Moral Frontier*. Malden, MA: Blackwell Publishing, 115.
136. Herman, L. M. (1980). Cognitive characteristics of dolphins. In L. M. Herman (Ed.), *Cetacean Behavior: Mechanisms and Functions* (pp. 363–429). New York: Wiley Interscience.
137. White, T. (2007). *In Defense of Dolphins: The New Moral Frontier*. Malden, MA: Blackwell Publishing.
138. Fitch, W. T. (2010). *The Evolution of Language*. Cambridge: Cambridge University Press, 164.
139. Chen, M. K., Lakshminarayanan, V., & Santos, L. R. (2006). How basic are behavioral biases? Evidence from capuchin monkey trading behavior. *Journal of Political Economy, 114*(3), 517–37.
140. Herman, L. M. (2009). Can dolphins understand language? In P. Sutcliffe, L. M. Stanford, & A. R. Lommel (Eds.), *LACUS Forum XXXIV: Speech and Beyond* (pp. 3–20). Houston, TX: LACUS, 7.
141. Murayama, T., Fujii, Y., Hashimoto, T., Shimoda, A., Iijima, S., Hayasaka, K., Shiroma, N., Koshikawa, M., Katsumata, H., Soichi, M., & Arai, K. (2012). Preliminary study of object labeling using sound production in a beluga. *International Journal of Comparative Psychology, 25*, 195–207.
142. Lilly, J. C. (1967). *The Mind of the Dolphin: A Nonhuman Intelligence*. New York: Doubleday.
143. Hooper J. (1983, January). John Lilly: altered states. *Omni Magazine*. A similar experiment was conducted at the Pt. Mugu Cetacean Facility in the 1960s that led to the following US Naval report (but no peer-reviewed publications): Batteau, D. W., & Markey, P. R. (1967). Man/dolphin communication. Final report prepared for U.S. Naval Ordance Test Station, China Lake, CA.
144. Reiss, D. (2011). *The Dolphin in the Mirror: Exploring Dolphin Minds and Saving Dolphin Lives*. Boston, MA: Houghton Mifflin Harcourt.
145. Hooper, S., Reiss, D., Carter, M., & McCowan, B. (2006). Importance of contextual saliency on vocal imitation by bottlenose dolphins. *International Journal of Comparative Psychology, 19*, 116–28.
146. Sigurdson, J. (1993). Frequency-modulated whistles as a medium for communication with the bottlenose dolphin. In H. L. Roitblat, L. M. Herman, &

P. E. Nachtigall (Eds.), *Language and Communication: Comparative Perspectives* (pp. 153–73). Hillsdale, NJ: Lawrence Erlbaum Associates.

147. Richards, D. G., Woltz, J. P., & Herman, L. M. (1984). Vocal mimicry of computer-generated sounds and labeling of objects by a bottlenose dolphin (*Tursiops truncatus*). *Journal of Comparative Psychology, 98*, 10–28.
148. Obituary for Kenneth Lee Marten, PhD (2010). *Aquatic Mammals, 36*(3), 323–5.
149. Xitco, M. J., Jr., Gory, J. D., & Kuczaj, S. A., II (1991). An introduction to The Living Seas' dolphin keyboard communication system. Presented at the 19th Annual Conference of the International Marine Animal Trainers Association, October, Concord, CA.
150. Herzing, D. L. (2010). SETI meets a social intelligence: dolphins as a model for real-time interaction and communication with a sentient species. *Acta Astronautica, 67*, 1451–4.
151. Herzing, D. (2011). *Dolphins Diaries: My 20 Years with Spotted Dolphins in the Bahamas.* New York: St. Martin's Press.
152. Campbell, M. (2011). Talk with a dolphin via underwater translation machine. *New Scientist, 2811.*
153. Lakatos, G., Gácsi, M., Topál, J., Miklósi, Á. (2012). Comprehension and utilisation of pointing gestures and gazing in dog–human communication in relatively complex situations. *Animal Cognition*, 15, 201–13.
154. Miklósi, A. (2009). Evolutionary approach to communication between humans and dogs. *Veterinary Research Communications, 33*(Suppl. 1), 53–9.
155. Herzing, D. L., Delfour, F. and Pack, A. A. (2012). Responses of human-habituated wild Atlantic spotted dolphins to play behaviors using a two-way interface. *International Journal of Comparative Psychology* 25(2), 137–65.
156. Markov, V. I., & Ostrovskaya, V. M. (1990). Organization of communication system in *Tursiops truncatus* Montagu. In J. A. Thomas & R. A. Kastelein (Eds.), *Sensory Abilities of Cetaceans—Laboratory and Field Evidence* (pp. 599–602). NATO ASI Series, Series A: Life Sciences, Vol. 196. New York: Plenum Press.
157. See, for example, this interview with Lori Marino at bigthink.com: Can dolphins understand humans? (February 11, 2010). Retrieved from http://bigthink.com/ideas/18648.
158. Shannon, C. E. (1948). A mathematical theory of communication. *Bell System Technical Journal, 27*, 379–423, 623–56.
159. Dreher, J. J. (1966). Cetacean communication: small group experiment. In K. Norris (Ed.), *Whales, Dolphins and Porpoises* (pp. 529–43). Berkeley and Los Angeles: University of California Press.
160. Tavolga, W. N. (Ed.) (1965). Technical Report: NAVTRADEVCEN 1212–1, Review of Marine Bio-Acoustics, State of the Art: 1964. (Port Washington, NY: US NAVAL Training Device Center, February 1965), 57.
161. Note that scientists have been using information theory for decades to study birdsong. For example, a study found 6.7 bits of information in chickadee calls (compared to 11.8 bits per word in English) using Shannon entropy calculations. See Hailman, J. P., Ficken, M. S., & Ficken, R. W. (1985). The "chick-a-dee"

calls of *Parus atricapillus*: a recombinant system of animal communication compared with written English. *Semiotica 56*, 191–224.

162. Markov, V. I., & Ostrovskaya, V. M. (1990). Organization of communication system in *Tursiops truncatus* Montagu. In J. A. Thomas & R. A. Kastelein (Eds.), *Sensory Abilities of Cetaceans—Laboratory and Field Evidence* (pp. 599–602). NATO ASI Series, Series A: Life Sciences, Vol. 196. New York: Plenum Press.
163. Vladimir I. Markov should not be confused with Andreyevich Markov, the Russian mathematician who invented Markov chains, which play a fundamental role in the mathematics of information theory.
164. McCowan, B., Hanser, S. F., & Doyle, L. R. (1999). Quantitative tools for comparing animal communication systems: information theory applied to bottlenose dolphin whistle repertoires. *Animal Behaviour, 57*, 409–19.
165. Suzuki, R., Buck, J. R., & Tyack, P. L. (2005). The use of Zipf's law in animal communication analysis. *Animal Behaviour, 69*(1), F9–F17.
166. Ferrer-i-Cancho, R., & McCowan, B. (2009). A law of word meaning in dolphin whistle types. *Entropy, 11*(4), 688–701.
167. McCowan, B., Doyle, L. R., Jenkins, J., & Hanser, S. F. (2005). The appropriate use of Zipf's law in animal communication studies. *Animal Behaviour, 69*. Pages F1 and F3 respectively.
168. Thanks to Mike Johnson for pointing out that "humans are fundamentally the best comprehensive pattern recognition systems there are, and for things like language you'd really expect people to do a better job of identifying structure and meaning" (from an email from Mike on May 12, 2011).
169. Lilly, J. C. (1978). *Communication Between Man and Dolphin: The Possibilities of Talking with Other Species*. New York: Crown Publishers, 156.
170. Global Heart, Inc. (2011). The discovery of dolphin language [Press Release]. Retrieved from http://www.speakdolphin.com/ResearchItems.cfm?ID=20.
171. For a thorough introduction to the principles of echolocation, see Au, W. W. L. (1993). *The Sonar of Dolphins*. New York: Springer-Verlag.
172. Aubauer, R., & Au, W. W. (1998). Phantom echo generation: a new technique for investigating dolphin echolocation. *Journal of the Acoustical Society of America, 104*(3 Pt 1), 1165–70.
173. deLong, C. M., Au, W. W. L., & Harley, H. E. (2002). Features of echoes bottlenose dolphins use to perceive object properties. *Journal of the Acoustical Society of America, 112*(5), 2335.
174. Houser, D., Martin, S. W., Bauer, E. J., Phillips, M., Herrin, T., Cross, M., Vidal, A., et al. (2005). Echolocation characteristics of free-swimming bottlenose dolphins during object detection and identification. *Journal of the Acoustical Society of America, 117*(4 Pt 1), 2308–17.
175. Houser, D. S., Helweg, D. A., & Moore, P. W. (1999). Classification of dolphin echolocation clicks by energy and frequency distributions. *Journal of the Acoustical Society of America, 106*(3 Pt 1), 1579–85.

176. Moore, P. W., Dankiewicz, L. A., and Houser, D. S. (2008). Beamwidth control and angular target detection in an echolocating bottlenose dolphin (*Tursiops truncatus*). *Journal of the Acoustical Society of America, 124*, 3324–32.
177. Muller, M. W., Allen, J. S., Au, W. W. L., & Nachtigall, P. E. (2008). Time-frequency analysis and modeling of the backscatter of categorized dolphin echolocation clicks for target discrimination. *Journal of the Acoustical Society of America, 124*(1), 657–66.
178. Helweg, D. A., Moore, P. W., Dankiewicz, L. A., Zafran, J. M., & Brill, R. L. (2003). Discrimination of complex synthetic echoes by an echolocating bottlenose dolphin. *Journal of the Acoustical Society of America, 113*(2), 1138–44.
179. Xitco, M. J., & Roitblat, H. L. (1996). Object recognition through eavesdropping: passive echolocation in bottlenose dolphins. *Animal Learning Behavior, 24*(4), 355–65.
180. Gregg, J. D., Dudzinski, K. M., & Smith, H. V. (2007). Do dolphins eavesdrop on the echolocation signals of conspecifics? *International Journal of Comparative Psychology, 20*, 65–88.
181. These are direct quotes from email exchanges I've had with people who have written to me via the Dolphin Communication Project.
182. Attributed to Bertrand Russell in an article titled "Is there a God?" Commissioned (but not published) by *Illustrated* magazine. Retrieved from http://cfpf.org.uk/articles/religion/br/br-god.html.
183. See, for example, the arguments for the use of the term "language" to describe ACS in Slobodchikoff, C. (2012). *Chasing Doctor Dolittle: Learning the Language of Animals*. New York: St. Martin's Press. Note however that the definition of language used in this book does not include the all important aspect of "limitless expression," nor place much emphasis on social cognitive aptitude as it pertains to theory of mind and mindreading. This results in nearly all forms of ACS falling under Slobodchikoff's definition of "language."
184. For an overview of research on dolphin communication, see Dudzinski, K., & Frohoff, T. (2008). *Dolphin Mysteries: Unlocking the Secrets of Communication*. New Haven, CT: Yale University Press.

Chapter 6

1. From the theme song to the television show *Flipper*, with lyrics by William Dunham.
2. For an overview of Lilly's arguments, see Burnett, D. G. (2010). A Mind in the Water. *Orion Magazine*.
3. Sandoz-Merrill, B. (2005). *In the Presence of High Beings: What Dolphins Want You to Know*. San Francisco: Council Oak Books, 19.
4. Keim, B. (July 19, 2012). New science emboldens long shot bid for dolphin, whale rights. *Wired Science*. Retrieved from http://www.wired.com/wired-science/ 2012/07/cetacean-rights.
5. PETA sues SeaWorld for violating orcas' constitutional rights (October 25, 2011). Retrieved from http://www.peta.org/b/thepetafiles/archive/2011/10/25/peta-sues-seaworld-for-violating-orcas-constitutional-rights.aspx.

6. Melillo, K. E., Dudzinski, K. M., & Cornick, L. A. (2009). Interactions between Atlantic spotted (*Stenella frontalis*) and bottlenose (*Tursiops truncatus*) dolphins off Bimini, The Bahamas, 2003–2007. *Aquatic Mammals, 35*(2), 281–91.
7. Dudzinski, K. M., Gregg, J. D., Melillo-Sweeting, K., Seay, B., Levengood, A., & Kuczaj, S. A. (2012). Tactile contact exchanges between dolphins: self-rubbing versus inter-individual contact in three species from three geographies. *International Journal of Comparative Psychology, 25,* 21–43.
8. Dudzinski, K. M., Gregg, J. D., Paulos, R. D., & Kuczaj, S. A. (2010). A comparison of pectoral fin contact behaviour for three distinct dolphin populations. *Behavioural Processes, 84,* 559–67.
9. Dudzinski, K. M., Gregg, J. D., Ribic, C. A., & Kuczaj, S. A. (2009). A comparison of pectoral fin contact between two different wild dolphin populations. *Behavioural Processes, 80,* 182–90.
10. Frohoff, T. G., & Peterson, B. (Eds.) (2003). *Between Species: Celebrating the Dolphin-Human Bond.* San Francisco, CA: Sierra Club Books.
11. See, for example, the gentle tactile exchanges accompanying courtship in slugs: Reise, H. (2006). A review of mating behavior in slugs of the genus Deroceras (Pulmonata: Agriolimacidae). *American Malacological Bulletin,* 23, 137–56.
12. Sugiyama, Y. (1988). Grooming interactions among adult chimpanzees at Bossou, Guinea, with special reference to social structure. *International Journal of Primatology, 9*(5), 393–407.
13. Maple, T., & Westlund, B. (1975). The integration of social interactions between cebus and spider monkeys in captivity. *Applied Animal Ethology, 1*(3), 305–8.
14. Tenow, O., Fagerström, T., & Wallin, L. (2008). Epimeletic behaviour in airborne common swifts *Apus apus*: do adults support young in flight? *Ornis Svecica 18,* 96–107.
15. Atkins, P. (Director), McCarey, K. (Writer). (1999). Dolphins: *The Dark Side* [Motion picture]. *National Geographic.*
16. Quote taken from National Geographic overview of Dolphins: *The Dark Side.* Retrieved from http://channel.nationalgeographic.com/wild/episodes/dolphins-the-dark-side1/.
17. Callen, K., & Cochrane, A. (1992). *Dolphins and Their Power to Heal.* Rochester, VI: Healing Arts Press, 67.
18. Callen, K., & Cochrane, A. (1992). *Dolphins and Their Power to Heal.* Rochester, VI: Healing Arts Press, 69.
19. For examples see Sandoz-Merrill, B. (2005). *In the Presence of High Beings: What Dolphins Want You to Know.* San Francisco, CA: Council Oak Books, 23 and 37.
20. Sandoz-Merrill, B. (2005). *In the Presence of High Beings: What Dolphins Want You to Know.* San Francisco, CA: Council Oak Books, 23.
21. Sirius Institute FAQ (April 24, 2005). Retrieved from http://www.planetpuna.com/faq.htm.
22. Killer dolphins slaying, sexually assaulting porpoises in San Francisco (November 21, 2011). *Huffington Post.* Retrieved from http://www.huffingtonpost.com/2011/11/21/dolphins-kill-porpoises_n_1106471.html.

23. Cotter, M. P., Maldini, D., & Jefferson, T. A. (2011). "Porpicide" in California: killing of harbor porpoises (*Phocoena phocoena*) by coastal bottlenose dolphins (*Tursiops truncatus*). *Marine Mammal Science, 28*(1), E1–E15.
24. Ross, H. M., & Wilson, B. (1996). Violent interactions between bottlenose dolphins and harbour porpoises. *Proceedings of the Royal Society of London—Series B: Biological Sciences, 263*(1368), 283–6.
25. Ross, H. M., & Wilson, B. (1996). Violent interactions between bottlenose dolphins and harbour porpoises. *Proceedings of the Royal Society of London—Series B: Biological Sciences, 263*(1368), 283–6. Pg. 283.
26. Ross, H. M., & Wilson, B. (1996). Violent interactions between bottlenose dolphins and harbour porpoises. *Proceedings of the Royal Society of London—Series B: Biological Sciences, 263*(1368), 283–6. Pg. 286.
27. Cotter, M. P., Maldini, D., & Jefferson, T. A. (2011). "Porpicide" in California: killing of harbor porpoises (*Phocoena phocoena*) by coastal bottlenose dolphins (*Tursiops truncatus*). *Marine Mammal Science, 28*(1), E1–E15.
28. As described in Cotter, M. P., Maldini, D., & Jefferson, T. A. (2011). "Porpicide" in California: killing of harbor porpoises (*Phocoena phocoena*) by coastal bottlenose dolphins (*Tursiops truncatus*). *Marine Mammal Science, 28*(1), E1–E15.
29. Wilkin, S. M., Cordaro, J., Gulland, F. M. D., Wheeler, E., Dunkin, R., Sigler, T., Casper, D., Berman, M., Flannery, M., Fire, S., Wang, Z., Colegrove, K., & Baker, J. (2012). An unusual mortality event of harbor porpoises (*Phocoena phocoena*) off central California: increase in blunt trauma rather than an epizootic. *Aquatic Mammals, 38*(3), 301–10.
30. Patterson, I. A., Reid, R. J., Wilson, B., Grellier, K., Ross, H. M., & Thompson, P. M. (1998). Evidence for infanticide in bottlenose dolphins: an explanation for violent interactions with harbour porpoises. *Proceedings of the Royal Society of London B: Biological Sciences, 265*(1402), 1167–70.
31. Earthwatch scientists capture dolphin attack on camera. (2009, October). *Earthwatch Institute eNewsletter*. Retrieved from http://www.earthwatch.org/europe/newsroom/science/news-3-attack.html.
32. Patterson, I. A., Reid, R. J., Wilson, B., Grellier, K., Ross, H. M., & Thompson, P. M. (1998). Evidence for infanticide in bottlenose dolphins: an explanation for violent interactions with harbour porpoises. *Proceedings of the Royal Society of London B: Biological Sciences, 265*(1402), 1167–70.
33. Dunn, D. G., Barco, S., McLellan, W. A., & Pabst, D. A. (1998). Virginia Atlantic bottlenose dolphin (*Tursiops truncatus*) strandings: gross pathological findings in ten traumatic deaths. Abstract, Sixth Annual Atlantic Coastal Dolphin Conference, May, Sarasota, Florida.
34. Milius, S. (July 18, 1998). Infanticide reported in dolphins. *Science News*. Retrieved from http://www.sciencenews.org/sn_arc98/7_18_98/fob1.htm.
35. Packer, C., Scheel, D., & Pusey, A. E. (1990). Why lions form groups: food is not enough. *American Naturalist, 136*, 1–19.
36. Scott, E. M., Mann, J., Watson, J. J., Sargeant, B. L., & Connor, R. C. (2005). Aggression in bottlenose dolphins: evidence for sexual coercion, male-male competition,

and female tolerance through analysis of tooth-rake marks and behaviour. *Behaviour, 142*, 21–44.

37. Connor, R. C., Smolker, R. A., & Richards, A. F. (1992). Two levels of alliance formation among male bottlenose dolphins (*Tursiops* sp.). *Proceedings of the National Academy of Sciences, 89*, 987–90.
38. Parsons, K. M., Durban, J. W., & Claridge, D. E. (2003). Male-male aggression renders bottlenose dolphin (*Tursiops truncatus*) unconscious. *Aquatic Mammals*, 29(3), 360–2.
39. Wells, R. S. (1991). The role of long-term study in understanding the social structure of a bottlenose dolphin community. In K. Pryor & K. S. Norris (Eds.), *Dolphin Societies: Discoveries and Puzzles* (pp. 199–225). Berkeley: University of California Press.
40. Wells, R. S., Scott, M. D., & Irvine, A. B. (1987). The ocial structure of free-ranging bottlenose dolphins. In H. Genoways (Ed.), *Current Mammalogy*, Vol. 1 (pp. 247–305). New York: Plenum Press.
41. Connor, R. C., Richards, A. F., Smolker, R. A., & Mann, J. (1996). Patterns of female attractiveness in Indian Ocean bottlenose dolphins. *Behaviour, 133*, 37–69.
42. Gibson, Q. A., & Mann, J. (2008). The size and composition of wild bottlenose dolphin (*Tursiops* sp.) mother-calf groups in Shark Bay, Australia. *Animal Behaviour, 76*, 389–405.
43. Lusseau, D., Schneider, K., Boisseau, O. J., Haase, P., Slooten, E., & Dawson, S. M. (2003). The bottlenose dolphin community of Doubtful Sound features a large proportion of long-lasting associations—can geographic isolation explain this unique trait? *Behavioral Ecology and Sociobiology, 54*, 396–405.
44. Lusseau, D. (2007). Why are male social relationships complex in the Doubtful Sound bottlenose dolphin population? *PLoS ONE 2*(4), e348.
45. Wilson, D. R. B. (1995). The ecology of bottlenose dolphins in the Moray Firth, Scotland: a population at the northern extreme of the species' range. PhD thesis, Aberdeen University, Aberdeen, UK.
46. Foley, A., McGrath, D., Berrow, S., & Gerritsen, H. (2010). Social structure within the bottlenose dolphin (*Tursiops truncatus*) population in the Shannon Estuary. *Aquatic Mammals 36*(4), 372–81.
47. Scott, E. M., Mann, J., Watson, J. J., Sargeant, B. L., & Connor, R. C. (2005). Aggression in bottlenose dolphins: evidence for sexual coercion, male-male competition, and female tolerance through analysis of tooth-rake marks and behaviour. *Behaviour, 142*, 21–44.
48. Hill, H. M., Greer, T., Solangi, M., & Kuczaj, S. A., II (2007). All mothers are not the same: maternal styles in bottlenose dolphins (*Tursiops truncatus*). *International Journal of Comparative Psychology, 20*, 34–53.
49. Pryor, K., & Kang-Shallenberger, I. (1991). Social structure in spotted dolphins (*Stenella attenuata*) in the tuna purse seine fishery in the eastern Tropical Pacific. In K. Pryor & K. S. Norris (Eds.), *Dolphin Societies: Discoveries and Puzzles* (pp. 161–98). Berkeley: University of California Press.
50. Wells, R. S., Scott M. D., & Blair Irvine, A. (1987). The social structure of free-ranging bottlenose dolphins. In H. H. Genoways (Ed.), *Current Mammalogy*, Vol. 1 (pp. 247–305). New York: Plenum Press.

51. McCowan, B., & Reiss, D. (1995). Maternal aggressive contact vocalizations in captive bottlenose dolphins (*Tursiops truncatus*): wide-band, low frequency signals during mother/aunt-infant interactions. *Zoo Biology, 14*(4), 293–310.
52. For an overview, see Frantzis, A., & Herzing, D. L. (2002). Mixed species associations of striped dolphin (*Stenella coeruleoalba*), short-beaked common dolphin (*Delphinus delphis*) and Risso's dolphin (*Grampus griseus*), in the Gulf of Corinth (Greece, Mediterranean Sea). *Aquatic Mammals, 28*(2), 188–97.
53. Deakos, M. H., Branstetter, B. K., Mazzuca, L., Fertl, D., & Mobley, J. R., Jr. (2010). Two unusual interactions between a bottlenose dolphin (*Tursiops truncatus*) and a humpback whale (*Megaptera novaeangliae*) in Hawaiian waters. *Aquatic Mammals, 36*(2), 121–8.
54. Kiszka, J. (2007). Atypical associations between dugongs (*Dugong dugon*) and dolphins in a tropical lagoon. *Journal of the Marine Biological Association UK, 87*, 101–4.
55. Wedekin, L. L., Daura-Jorge, F. G., & Simões-Lopes, P. C. (2004). An aggressive interaction between bottlenose dolphins (*Tursiops truncatus*) and estuarine dolphins (*Sotalia guianensis*) in Southern Brazil. *Aquatic Mammals, 30*(3), 391–7.
56. Herzing, D. L., & Johnson, C. M. (1997). Interspecific interactions between Atlantic spotted dolphins (*Stenella frontalis*) and bottlenose dolphins (*Tursiops truncatus*) in the Bahamas, 1985–1995. *Aquatic Mammals 23*(2), 85–100.
57. Melillo, K. E., Dudzinski, K. M., & Cornick, L. A. (2009). Interactions between Atlantic spotted (*Stenella frontalis*) and bottlenose (*Tursiops truncatus*) dolphins off Bimini, The Bahamas, 2003–2007. *Aquatic Mammals, 35*(2), 281–91.
58. Frantzis, A., & Herzing, D. L. (2002). Mixed species associations of striped dolphin (*Stenella coeruleoalba*), short-beaked common dolphin (*Delphinus delphis*) and Risso's dolphin (*Grampus griseus*), in the Gulf of Corinth (Greece, Mediterranean Sea). *Aquatic Mammals, 28*(2), 188–97.
59. Haelters, J., & Everaarts, E. (2011). Two cases of physical interaction between white-beaked dolphins (*Lagenorhynchus albirostris*) and juvenile harbour porpoises (*Phocoena phocoena*) in the southern North Sea. *Aquatic Mammals, 37*(2), 198–201.
60. Baird, R. W. (1998). An interaction between Pacific white-sided dolphins and a neonatal harbor porpoise. *Mammalia, 62*(1), 134–9.
61. Coscarella, M. A., & Crespo, E. A. (2009). Feeding aggregation and aggressive interaction between bottlenose (*Tursiops truncatus*) and Commerson's dolphins (*Cephalorhynchus commersonii*) in Patagonia, Argentina. *Journal of Ethology, 28*(1), 183–7.
62. Barnett, J., Davison, N., Deaville, R., Monies, R., Loveridge, J., Tregenza, N., & Jepson, P. D. (2009). Postmortem evidence of interactions of bottlenose dolphins (*Tursiops truncatus*) with other dolphin species in south-west England. *Veterinary Record 165*(15), 441–4.
63. See Mooney, J. (1997). *Captive Cetaceans: A Handbook for Campaigners. A Report for the Whale and Dolphin Conservation Society*; or, White, T. (2007). *In Defense of Dolphins: The New Moral Frontier*. Malden, MA: Blackwell Publishing.

64. Geraci, J. (1984). Marine mammals. In *Guide to the Care and Use of Experimental Animals*, Vol. 2 (pp. 131–42). Ottawa: Canadian Council on Animal Care; and also Sweeney, J. (1990). Marine mammal behavioral diagnostics. In L. Dierauf (Ed.), *Handbook of Marine Mammal Medicine* (pp. 53–72). Boca Raton, FL: CRC Press.
65. Ostman, J. (1991). Changes in aggressive and sexual behavior between two male bottlenose dolphins (*Tursiops truncatus*) in a captive colony. In K. Pryor & K. S. Norris (Eds.), *Dolphin Societies: Discoveries and Puzzles* (pp. 305–18). Berkeley: University of California Press.
66. Pryor, K., & Kang-Shallenberger, I. (1991). Social structure in spotted dolphins (*Stenella attenuata*) in the tuna purse seine fishery in the eastern Tropical Pacific. In K. Pryor & K. S. Norris (Eds.), *Dolphin Societies: Discoveries and Puzzles* (pp. 161–98). Berkeley: University of California Press.
67. Sweeney, J. (1990). Marine mammal behavioral diagnostics. In L. Dierauf (Ed.), *Handbook of Marine Mammal Medicine* (pp. 53–72). Boca Raton: CRC Press.
68. Gareci, J. (1986). Husbandry. In M. Fowler (Ed.), *Zoo and Wild Animal Medicine* (pp. 757–60). Philadelphia, PA: Harcourt Brace Jovanovich.
69. Connor, R. C., Wells, R. S., Mann, J., & Read, A. J. (2000). The bottlenose dolphin: social relationships in a fission-fusion society. In J. Mann, R. C. Connor, P. L. Tyack, & H. Whitehead (Eds.), *Cetacean Societies: Field Studies of Dolphins and Whales* (pp. 91–126). Chicago, IL: University of Chicago Press, 107.
70. Connor, R. C. (2000). Group living in whales and dolphins. In J. Mann, R. C. Connor, P. L. Tyack, & H. Whitehead (Eds.), *Cetacean Societies: Field Studies of Dolphins and Whales* (pp. 199–218). Chicago, IL: University of Chicago Press.
71. Performing whale dies in collision with another (August 23, 1989). *New York Times*. Retrieved from http://www.nytimes.com/1989/08/23/us/performing-whale-dies-in-collision-with-another.html.
72. Most aggression witnessed in spotted dolphins involves threat displays like open mouths/gaping: see Pryor, K., & Kang-Shallenberger, I. (1991). Social structure in spotted dolphins (*Stenella attenuata*) in the tuna purse seine fishery in the eastern Tropical Pacific. In K. Pryor & K. S. Norris (Eds.), *Dolphin Societies: Discoveries and Puzzles* (pp. 161–98). Berkeley: University of California Press.
73. Pryor, K. W. (1990). Non-acoustic communication in small cetaceans: glance, touch, position, gesture, and bubbles. In J. A. Thomas & R. Kastelein (Eds.), *Sensory Abilities of Cetaceans* (pp. 537–44). New York: Plenum Press.
74. Herman, L. M., & Tavolga, W. N. (1980). The communication systems of cetaceans. In L. M. Herman (Ed.), *Cetacean Behavior: Mechanisms and Functions* (pp. 149–209). New York: John Wiley & Sons.
75. Connor, R. C., Wells, R. S., Mann, J., & Read, A. J. (2000). The bottlenose dolphin: social relationships in a fission-fusion society. In J. Mann, R. C. Connor, P. L. Tyack, & H. Whitehead (Eds.), *Cetacean Societies: Field Studies of Dolphins and Whales* (pp. 91–126). Chicago, IL: University of Chicago Press.
76. Dudzinski, K. M. (1998). Contact behavior and signal exchange in Atlantic spotted dolphins(*Stenella frontalis*). *Aquatic Mammals, 24*(3), 129–42.

77. Dudzinski, K. M., Thomas, J., & Gregg, J. D. (2009). Communication. In W. F. Perrin, B. Würsig, & H. C. M. Thewissen (Eds.), *Encyclopedia of Marine Mammals*, 2nd edn (pp. 260–58). New York: Academic Press.
78. Connor, R. C., Wells, R. S., Mann, J., & Read, A. J. (2000). The bottlenose dolphin: social relationships in a fission-fusion society. In J. Mann, R. C. Connor, P. L. Tyack, & H. Whitehead (Eds.), *Cetacean Societies: Field Studies of Dolphins and Whales* (pp. 91–126). Chicago, IL: University of Chicago Press.
79. Frantzis, A., & Herzing, D. L. (2002). Mixed species associations of striped dolphin (*Stenella coeruleoalba*), short-beaked common dolphin (*Delphinus delphis*) and Risso's dolphin (*Grampus griseus*), in the Gulf of Corinth (Greece, Mediterranean Sea). *Aquatic Mammals, 28*(2), 188–97.
80. Scott, E. M., Mann, J., Watson, J. J., Sargeant, B. L., & Connor, R. C. (2005). Aggression in bottlenose dolphins: evidence for sexual coercion, male-male competition, and female tolerance through analysis of tooth-rake marks and behaviour. *Behaviour, 142*, 21–44.
81. MacLeod, C. D. (1998). Intraspecific scarring in odontocete cetaceans: an indicator of male "quality" in aggressive social interactions. *Journal of Zoology*, 244, 71–7.
82. Martin, A. R., & Da Silva, V. M. F. (2006). Sexual dimorphism and body scarring in the boto (Amazon river dolphin) *Inia Geoffrensis. Marine Mammal Science, 22*(1), 25–33.
83. Jenkins, M. (2009, June). River spirits. *National Geographic Magazine*, 98–111. Retrieved from http://ngm.nationalgeographic.com/2009/06/dolphins/jenkins-text.
84. Mann, J., & Smuts, B. (1999). Behavioral development in wild bottlenose dolphin newborns (*Tursiops* sp.). *Behaviour, 136*, 529–66.
85. von Streit, C., Udo Ganslosser, U., & von Fersen, L. (2011). Ethogram of two captive mother-calf dyads of bottlenose dolphins (*Tursiops truncatus*): comparison with field ethograms. *Aquatic Mammals, 37*(2), 193–7.
86. Miles J. A., & Herzing D. L. (2003). Underwater analysis of the behavioural development of free-ranging Atlantic spotted dolphin (*Stenella frontalis*) calves (birth to 4 years of age). *Aquatic Mammals, 29*(3), 363–77.
87. This ethogram was developed as part of the following study: Dudzinski, K. M. (1996). Communication and behavior in the Atlantic spotted dolphins (*Stenella frontalis*): relationship between vocal and behavioral activities. Dissertation Thesis, Texas A&M University, College Station.
88. Herzing, D. L. (1996). Vocalizations and associated underwater behavior of free-ranging Atlantic spotted dolphins, *Stenella frontalis*, and bottlenose dolphins, *Tursiops truncatus*. *Aquatic Mammals 22*(2), 61–79.
89. For example, Gaskin, D. E. (1972). *Whales, Dolphins and Seals, with Special Reference to the New Zealand Region*. Auckland: Heinemann Educational Books.
90. Visser, I. (1998). Prolific body scars and collapsing dorsal fins on killer whales (*Orcinus orca*) in New Zealand waters. *Aquatic Mammals, 24*(2), 71–81.
91. Connor, R. C., Wells, R. S., Mann, J., & Read, A. J. (2000). The bottlenose dolphin: social relationships in a fission-fusion society. In J. Mann, R. C. Connor, P. L. Tyack, & H. Whitehead (Eds.), *Cetacean Societies: Field Studies of Dolphins and Whales*, (pp. 91–126). Chicago, IL: University of Chicago Press, 102.

92. For an overview, see Connor, R. C. (2000). Group living in whales and dolphins. In J. Mann, R. C. Connor, P. L. Tyack, & H. Whitehead (Eds.), *Cetacean Societies: Field Studies of Dolphins and Whales* (pp. 199–218). Chicago, IL: University of Chicago Press.
93. Cantor, M., Wedekin, L. L., Guimaraes, P. R., Daura-Jorge, F. G. Rossi-Santos, M. R., & Simoes-Lopes, P. C. (2012). Disentangling social networks from spatiotemporal dynamics: the temporal structure of a dolphin society. *Animal Behaviour, 84*, 641–51.
94. Connor, R. C., Wells, R. S., Mann, J., & Read, A. J. (2000). The bottlenose dolphin: social relationships in a fission-fusion society. In J. Mann, R. C. Connor, P. L. Tyack, & H. Whitehead (Eds.), *Cetacean Societies: Field Studies of Dolphins and Whales* (pp. 91–126). Chicago, IL: University of Chicago Press, 91.
95. Archie, E. A., Moss, C. J., and Alberts, S. C. (2006). The ties that bind: genetic relatedness predicts the fission and fusion of social groups in wild African elephants. *Proceedings of the Royal Society B*, 273: 513–22.
96. van Schaik, C. P. (1999). The socioecology of fission-fusion sociality in orangutans. *Biomedical and Life Sciences, 40*(1), 69–86.
97. Connor R. C., & Vollmer, N. (2009). Sexual coercion in dolphin consortships: a comparison with chimpanzees. In M. N. Muller & R. W. Wrangham (Eds.), *Sexual Coercion in Primates: An Evolutionary Perspective on Male Aggression against Females* (pp. 218–43). Cambridge, MA: Harvard University Press.
98. Connor, R. C., Smolker, R. A., & Richards, A. F. (1992). Two levels of alliance formation among bottlenose dolphins (*Tursiops* sp.). *Proceedings of the National Academy of Sciences, 89*, 987–90.
99. Randic, S., Connor, R. C., Sherwin, W. B., & M. Krützen. (2012). A novel mammalian social structure in Indo-Pacific bottlenose dolphins (*Tursiops* sp.): complex male alliances in an open social network. *Proceedings of the Royal Society B: Biological Sciences*. doi:10.1098/rspb.2012.0264.
100. Connor, R. C., Heithaus, M. R., & Barre, L. M. (2001). Complex social structure, alliance stability and mating access in a bottlenose dolphin "super-alliance." *Proceedings of the Royal Society B: Biological Sciences, 268*, 263–7.
101. Connor, R. C., Watson-Capps, J., Sherwin, W. B., & Krutzen, M. (2010). A new level of complexity in the male alliance networks of Indian ocean bottlenose dolphins (*Tursiops* sp.). *Biology Letters, 6*(20), 1–4.
102. Connor, R. C. (2007). Dolphin social intelligence: complex alliance relationships in bottlenose dolphins and a consideration of selective environments for extreme brain size evolution in mammals. *Philosophical Transactions of the Royal Society B: Biological Sciences, 362*, 587–602.
103. Randic, S., Connor, R. C., Sherwin, W. B., & Krützen, M. (2012). A novel mammalian social structure in Indo-Pacific bottlenose dolphins (*Tursiops* sp.): complex male alliances in an open social network. *Proceedings of the Royal Society B: Biological Sciences*. doi:10.1098/rspb.2012.0264.
104. Baird, R. W. (2000). The killer whale—foraging specializations and group hunting. In Mann, R. C. Connor, P. L. Tyack, & H. Whitehead (Eds.), *Cetacean Societies: Field Studies of Dolphins and Whales* (pp. 127–53). Chicago, IL: University of Chicago Press.

105. Dozens of primate species engage in sex at times outside of estrus, which means they are technically having "sex for pleasure." For an overview, see Rice, S. A. (2007). *Encyclopedia of Evolution*. New York: Facts on File, Inc. Also, same-sex sexual behavior is common in the animal kingdom, which also, by definition, cannot result in egg fertilization, and therefore occurs only "for pleasure," rendering the idea that dolphins are one of the only animals to engage in sex for pleasure as demonstrably incorrect. More information on this topic can be found on my blog: http:www.justingregg.com.
106. There is an extremely long list of animal species that regularly engage in same-sex sexual behavior. More information on this subject in Bagemihl, B. (1999). *Biological Exuberance: Animal Homosexuality and Natural Diversity*. New York: St. Martin's Press.
107. The "blowhole sex" meme has been mentioned on *The Colbert Report* and *30 Rock* and been depicted in sculptures created by the artist Rune Olsen that have been displayed around the world. The idea originates from the following article describing something akin to dolphins inserting penises into blowholes: Renjun, L., Gewalt, W., Neurohr, B., & Winkler, A. (1994). Comparative studies on the behaviour of *Inia geoffrensis* and *Lipotes vexillifer* in artificial environments. *Aquatic Mammals, 20*(1), 39–45. However, when I queried one of the authors of this article as to whether or not it was the case that the dolphins truly did insert their penises into blowholes, this turned out not to have occurred. Nor has it ever been observed or mentioned in any other scientific literature. More information on this topic can be found on my blog: http:www.justingregg.com.
108. Dawkins, R. (May 24, 2012). The descent of Edward Wilson. *Prospect*. Retrieved from http://www.prospectmagazine.co.uk/magazine/edward-wilson-social-conquest-earth-evolutionary-errors-origin-species/.
109. Wilson, E. O. (2012). *The Social Conquest of Earth*. New York: Norton.
110. Conner, R. C. (1995). The benefits of mutualism: a conceptual framework. *Biological Reviews, 70*, 427–57.
111. Würsig, B., & Würsig, M. (Eds.) (2010). *The Dusky Dolphin: Master Acrobat off Different Shores*. San Diego, CA: Elsevier.
112. Vaughn, R. L., Muzi, E., Richardson, J. L., & Würsig, B. (2011). Dolphin bait-balling behaviors in relation to prey ball escape behaviors. *Ethology, 117*, 859–71.
113. Benoit-Bird, K. J., & Au, W. W. L. (2009). Cooperative prey herding by the pelagic dolphin, *Stenella longirostris*. *Journal of the Acoustical Society of America, 125*, 125–37.
114. Gazda, S. K, Connor, R. C., Edgar, R. K., & Cox, F. (2005). A division of labour with role specialization in group-hunting bottlenose dolphins (*Tursiops truncatus*) off Cedar Key, Florida. *Proceedings of the Royal Society B: Biological Sciences, 272*(1559), 135–40.
115. Duffy-Echevarria, E. E., Connor, R. C., & St. Aubin, D. J. (2008). Observations of strand-feeding behavior by bottlenose dolphins (*Tursiops truncatus*) in Bull Creek, South Carolina. *Marine Mammal Science, 24*, 202–6.

116. Daura-Jorge, F. G., Cantor, M., Ingram, S. N., Lusseau, D., & Simões-Lopes, P. C. (2012). The structure of a bottlenose dolphin society is coupled to a unique foraging cooperation with artisanal fishermen. *Biology Letters.* doi:10.1098/rsbl.2012.0174.
117. Neil, D. T. (2002). Cooperative fishing interactions between Aboriginal Australians and dolphins in eastern Australia. *Anthrozoös, 15,* 3–18.
118. Biju Kumar, A., Smrithy R., & Sathasivam (2012). Dolphin-assisted cast net fishery in the Ashtamudi Estuary, south-west coast of India. *Indian Journal of Fisheries, 59*(3), 143–8.
119. Busnel, R.G. (1973). Symbiotic relationship between man and dolphins. *Transactions of the New York Academy of Sciences, 35,* 112–31.
120. Connor, R. C., & Norris, K. S. (1982). *Are dolphins reciprocal altruists? American Naturalist,* 119, 358–74.
121. Baird, R. W., & Dill, L. M. (1997). Ecological and social determinants of group size in transient killer whales. *Behavioral Ecology 7*(4), 408–16.
122. Smith, T. G., Siniff, D. B., Reichle, R., & Stone, S. (1981). Coordinated behavior of killer whales, *Orcinus orca,* hunting a crabeater seal, *Lobodon carcinophagus. Canadian Journal of Zoology, 59,* 1185–9.
123. Visser, I. N., Smith, T. G., Bullock, I. D., Green, G. D., Carlsson, O. G. L., & Imberti, S. (2008). Antarctic peninsula killer whales (*Orcinus orca*) hunt seals and a penguin on floating ice. *Marine Mammal Science, 24,* 225–34.
124. Pitman, R. L., & Durban, J. W. (2012). Cooperative hunting behavior, prey selectivity and prey handling by pack ice killer whales (*Orcinus orca*), type B, in Antarctic Peninsula waters. *Marine Mammal Science, 28*(1), 16–36.
125. Würsig, B. (1986). Delphinid foraging strategies. In R. J. Schusterman, J. A. Thomas, & F. G. Wood (Eds.), *Dolphin Cognition and Behaviour: A Comparative Approach* (pp. 347–59). Hillsdale, NJ: Lawrence Erlbaum Associates.
126. Connor, R. C. (2000). Group living in whales and dolphins. In J. Mann, R. C. Connor, P. L. Tyack, & H. Whitehead (Eds.), *Cetacean Societies: Field Studies of Dolphins and Whales* (pp. 199–218). Chicago, IL: University of Chicago Press, 210.
127. Caldwell, M. C., & Caldwell, D. K. (1966). Epimeletic (care-giving) behavior in cetacea. In K. S. Norris (Ed.), *Whales, Dolphins and Porpoises* (pp. 755–89). Berkeley and Los Angeles: University of California Press.
128. Connor, R. C., & Norris, K. S. (1982). Are dolphins reciprocal altruists? *American Naturalist,* 119, 358–74.
129. Lee, P. C. (1986). Early social development among African elephant calves. *National Geographic Research, 2,* 388–401.
130. Pusey, A. (1983). Mother-offspring relationships in chimpanzees after weaning. *Animal Behaviour, 31*(2), 363–77.
131. Tyack, P. (2009). Behavior, overview. In W. F. Perrin, B. Würsig, & H. C. M. Thewissen (Eds.), *Encyclopedia of Marine Mammals,* 2nd edn (pp. 101–108). New York: Academic Press.
132. Grellier, K., Hammond, P. S., Wilson, B., Sanders-Reed, C. A., & Thompson, P. M. (2003). Use of photo-identification data to quantify mother–calf association patterns in bottlenose dolphins. *Canadian Journal of Zoology, 81,* 1421–7.

133. Cockcroft, V. G., & Ross, G. J. B. (1990). Observations on the early development of a captive bottlenose dolphin calf. In S. Leatherwood and R. R. Reeves (Eds.), *The Bottlenose Dolphin* (pp. 461–78). New York: Academic Press.
134. Ford, J. K. (2009). Killer whale. In W. F. Perrin, B. Würsig, & H. C. M. Thewissen (Eds.), *Encyclopedia of Marine Mammals*, 2nd edn (pp. 650–7). New York: Academic Press.
135. McAuliffe, K., & Whitehead, H. (2005). Eusociality, menopause and information in matrilineal whales. *Trends in Ecology & Evolution, 20*, 650.
136. Foster, E. A., Franks, D. W., Mazzi, S., Darden, S. K., Balcomb, K. C., Ford, J. K. B., & Croft, D. P. (2012). Adaptive prolonged postreproductive life span in killer whales. *Science, 337*(6100), 1313.
137. Wilson, E. O. (1975). *Sociobiology: The New Synthesis*. Cambridge, MA: Harvard University Press.
138. Bearzi, G. (1997). A "remnant" common dolphin observed in association with bottlenose dolphins in the Kvarneric (northern Adriatic Sea). *European Research on Cetaceans, 10*, 204.
139. Simard, P., & Gowans, S. (2004). Two calves in echelon: an alloparental association in Atlantic white-sided dolphins (*Lagenorhynchus acutus*). *Aquatic Mammals 30*(2), 330–4.
140. Mann, J., & Smuts, B. B. (1999). Behavioral development in wild bottlenose dolphin newborns (*Tursiops* sp.). *Behaviour, 136*, 529–66.
141. Karczmarski, L., Thornton, M., & Cockcroft, V. G. (1997). Description of selected behaviours of humpback dolphins *Sousa chinensis*. *Aquatic Mammals, 23*(3), 127–33.
142. Gaspar, C., Lenzi, R., Reddy, M. L., & Sweeney, J. (2000). Spontaneous lactation by an adult *Tursiops truncatus* in response to a stranded *Steno bredanensis* calf. *Marine Mammal Science, 16*, 653–8.
143. Huck, M., & Fernandez-Duque, E. (2012). When dads help: male behavioral care during primate infant development. In K. B. H. Clancy, K. Hinde, & J. N. Rutherford (Eds.), *Building Babies: Primate Development in Proximate and Ultimate Perspective* (pp. 361–85). New York: Springer.
144. Dudzinski, K. M., Gregg, J. D., Melillo-Sweeting, K., Seay, B., Levengood, A., & Kuczaj, S. A. (2012). Tactile contact exchanges between dolphins: self-rubbing versus inter-individual contact in three species from three geographies. *International Journal of Comparative Psychology, 25*, 21–43.
145. Warren-Smith, A. B., & Dunn, W. L. (2006). Epimeletic behaviour toward a seriously injured juvenile bottlenose dolphin (*Tursiops* sp.) in Port Phillip, Victoria, Australia. *Aquatic Mammals, 32*(3), 357–62.
146. Connor, R. C., & Norris, K. S. (1982). Are dolphins reciprocal altruists? *American Naturalist, 119*, 358–74.
147. Gibson, Q. A. (2006). Non-lethal shark attack on a bottlenose dolphin (*Tursiops* sp.) calf. *Marine Mammal Science, 22*(1), 192–8.
148. Mann, J., & H. Barnett. (1999). Lethal tiger shark (*Galeocerdo cuvieri*) attack on bottlenose dolphin (*Tursiops* sp.) calf: defense and reactions by the mother. *Marine Mammal Science, 15*(2), 568–75.

149. Wood, F. G., Caldwell, D. K., & Caldwell, M. C. (1970). Behavioral interactions between porpoises and sharks. *Investigations on Cetacea, 2*, 264–77.
150. Heithaus, M. R. (2001). Predator-prey and competitive interactions between sharks (order Selachii) and dolphins (suborder Odontoceti): a review. *Journal of Zoology, 253*, 53–68.
151. Saayman, G. S., & Tayler, C. K. (1979). The socioecology of humpback dolphins (*Sousa* spp.). In H. E. Winn & B. L. Olla (Eds.), *Behaviour of Marine Animals*, Vol. 3: *Cetaceans* (pp. 165–226). New York: Plenum Press.
152. Connor, R. C., & Norris, K. S. (1982). Are dolphins reciprocal altruists? *American Naturalist*, 119, 358–74.
153. Santos, M. C. O., Rosso, S., Siciliano, S., Zerbini, A., Zampirolli, E., Vicente, A. F., & Alvarenga, F. (2000). Behavioral observations of the marine tucuxi dolphin (*Sotalia fluviatilis*) in Sao Paulo estuarine waters, Southeastern Brazil. *Aquatic Mammals, 26*(3), 260–7.
154. Cockcroft, V. G., & Sauer, W. (1990). Observed and inferred epimeletic (nurturant) behaviour in bottlenose dolphins. *Aquatic Mammals, 16*(1), 31–2.
155. Fertl D., & Schiro A. (1994). Carrying of dead calves by free-ranging Texas bottlenose dolphins (*Tursiops truncatus*). *Aquatic Mammals, 20*(1), 53–6.
156. Ritter, F. (2007). Behavioral responses of rough-toothed dolphins to a dead newborn calf. *Marine Mammal Science, 23*(2), 429–33.
157. Ritter, F. (2002). Behavioral observations of rough-toothed dolphins (*Steno bredanensis*) off La Gomera, Canary Islands (1995–2000), with special reference to their interactions with humans. *Aquatic Mammals, 28*(1), 46–59.
158. Harzen. S., & dos Santos. M. E. (1992). Three encounters with wild bottlenose dolphins (*Tursiops truncatus*) carrying dead calves. *Aquatic Mammals*, 18(2), 49–55.
159. Lodi, L. (1992). Epimeletic behavior of free-ranging rough-toothed dolphins, *Steno bredanensis*, from Brazil. *Marine Mammal Science, 8*, 284–7.
160. Dudzinski, K. M., Sakai, M., Masaki, K., Kogi, K., Hishii, T., & Kurimoto, M. (2003). Behavioral observations of adult and sub-adult dolphins towards two dead bottlenose dolphins (one female and one male). *Aquatic Mammals*, 29(1), 108–16.
161. Connor, R. C., & Norris, K. S. (1982). Are dolphins reciprocal altruists? *American Naturalist, 119*, 358–74.
162. Perrin, W. F., & Geraci, J. R. (2009). Stranding. In W. F. Perrin, B. Würsig, & H. C. M. Thewissen (Eds.). *Encyclopedia of Marine Mammals*, 2nd edn (pp. 118–23). New York: Academic Press.
163. Packer, C., & Ruttan, L. (1988). The evolution of cooperative hunting. *American Naturalist, 132*(2), 159–98.
164. See overview in Fertl, D., & Schiro, A. (1994). Carrying of dead calves by free-ranging Texas bottlenose dolphins (*Tursiops truncatus*). *Aquatic Mammals, 20*(1), 53–6.
165. Conover, M. R. (1987). Acquisition of predator information by active and passive mobbers in ring-billed gull colonies. *Behaviour*, 102, 41–57.
166. Graw B., & Manser M. B. (2007). The function of mobbing in cooperative meerkats. *Animal Behaviour, 74*, 507–17.

167. Shields, W. M. (1984). Barn swallow mobbing: self-defence, collateral kin defence, group defence, or parental care? *Animal Behaviour*, 32, 132–48.
168. McLean, I. G., Smith, J. N. M., & Stewart, K. G. (1986). Mobbing behaviour, nest exposure, and breeding success in the American robin. *Behaviour*, 96, 171–86.
169. Passamani, M. (1995). Field observation of a group of Geoffroys marmosets mobbing a margay cat. *Folia Primatologica, 64*, 163–6.
170. Coss, R. G., & Biardi, J. E. (1997). Individual variation in the antisnake behavior of California ground squirrels (*Spermophilus beecheyi*). *Journal of Mammalogy, 78*(2), 294–310.
171. Helfman, G. S. (1989). Threat-sensitive predator avoidance in damselfish-trumpetfish interactions. *Behavioral Ecology and Sociobiology*, 24, 47–58.
172. Dominey, W. J. (1983). Mobbing in colonially nesting fishes, especially the bluegill, *Lepomis macrochirus*. *Copeia, 1983*(4), 1086–8.
173. Ratnieks, F. L. W., & Helantera, H. (2009). The evolution of extreme altruism and inequality in insect societies. *Philosophical Transactions of the Royal Society B: Biological Sciences*, 364(1533), 3169.
174. Bshary, R., Hohner, A., Ait-el-Djoudi, K., & Fricke, H. (2006). Interspecific communicative and coordinated hunting between groupers and giant moray Eels in the Red Sea. *PLoS Biology* 4(12), e431. doi:10.1371/journal.pbio.0040431.
175. Tenow, O., Fagerström, T., & Wallin, L. (2008). Epimeletic behaviour in airborne common swifts *Apus apus*: do adults support young in flight? *Ornis Svecica 18*, 96–107.
176. Rivas, J. A., & Levín, L. E. (2004). Sexually dimorphic anti-predator behavior in juvenile green iguanas *Iguana iguana*: evidence for kin selection in the form of fraternal care. In A. C. Alberts, R. L Carter, W. K. Hayes, & E. P Martins (Eds.), *Iguanas: Biology and Conservation* (pp. 119–26). Berkeley: University of California Press.
177. Bekoff, M. (2000). Animal emotions: exploring passionate natures. *BioScience*, 50(10), 861–70.
178. Norris, K. S., & Prescott, J. H. (1961). Observations on Pacific cetaceans of Californian and Mexican waters. *University of California Publications in Zoology, 63*, 291–402.
179. Shane, S. H. (1994). Pilot whales carrying dead sea lions. *Mammalia, 58*, 494–8.
180. Harzen. S., & dos Santos. M. E. (1992). Three encounters with wild bottlenose dolphins (*Tursiops truncatus*) carrying dead calves. *Aquatic Mammals, 18*(2), 49–55. Pg 55.
181. Griffin, D. R. (1984). *Animal Thinking*. Cambridge, MA: Harvard University Press.
182. Caldwell, M. C., & Caldwell, D. K. (1964). Experimental studies on factors involving care-giving behavior in three species of the cetacean family Delphinidae. *Bulletin of the Southern Californian Academy of Sciences, 63*(1), 1–20.
183. Iglesias, T. L, McElreath, R., & Patricelli, G. L. (2012). Western scrub-jay funerals: cacophonous aggregations in response to dead conspecifics *Animal Behaviour*. doi:10.1016/j.anbehav.2012.08.007.
184. Dudzinski, K. M., Sakai, M., Masaki, K., Kogi, K., Hishii, T., & Kurimoto, M. (2003). Behavioral observations of adult and sub-adult dolphins towards two dead bottlenose dolphins (one female and one male). *Aquatic Mammals, 29*(1), 108–16.

185. White, T. (2007). *In Defense of Dolphins: The New Moral Frontier*. Malden, MA: Blackwell Publishing.
186. Frohoff, T. (2000). The dolphin's smile. In M. Bekoff (Ed.), *The Smile of a Dolphin: Remarkable Accounts of Animal Emotions* (pp. 78–9). London: Discovery Books, 79.
187. For example, see discussion by McCulloch, D. (1999). Synchronicity: the dance of the dolphins. *Dolphin Synergy*. Retrieved from http://www.dolphinsynergy.com/lore.html.
188. This occurred at the Dolphin Communication Project's research site in Mikura, Japan.

Chapter 7

1. Grandin, T., & Johnson, C. (2006). *Animals in Translation*. London: Bloomsbury Publishing, 303.
2. Note that false killer whales appear to fail the test, and killer whale results are equivocal: Delfour, F., & Marten, K. (2001). Mirror image processing in three marine mammal species: Killer whales (*Orcinus orca*), false killer whales (*Pseudorca crassidens*) and California sea lions (*Zalophus californianus*). *Behavioural Processes, 53*, 181–90.
3. Cheke, L. G., Bird, C. D., & Clayton, N. S. (2011). Tool-use and instrumental learning in the Eurasian jay (*Garrulus glandarius*). *Animal Cognition*, 14(3), 441–55.
4. Vonk, J., & Beran, M. J. (2012). Bears "count" too: quantity estimation and comparison in black bears, *Ursus americanus*. *Animal Behaviour, 84*(1), 231–8.
5. Inoue, S., & Matsuzawa, T. (2007). Working memory of numerals in chimpanzees. *Current Biology, 17*, R1004–R1005.
6. Mora, C. V., Davison, M., Wild, J. M., and Walker, M. M. (2004). Magnetoreception and its trigeminal mediation in the homing pigeon. *Nature*, 432, 508–11.
7. Bentham, J. (1907). *An Introduction to the Principles of Morals and Legislation*. Oxford: Clarendon Press.
8. Singer, P. (1975). *Animal Liberation: A New Ethics for our Treatment of Animals*. New York: Random House. Note however that in later writings, Singer introduces the idea of self-consciousness as being a vital criterion for moral consideration. For example: Singer, P. (1979). *Practical Ethics*. Cambridge: Cambridge University Press.
9. Francione, G. L. (2008). *Animals as Persons: Essays on the Abolition of Animal Exploitation*. New York: Columbia University Press.
10. Thanks to Thomas White for email exchanges to help clarify his position on this point.
11. See, for example, Dawkins, M. S. (2001). Who needs consciousness? *Animal Welfare 10*, S19–S29.
12. Consider this quote from cetacean expert Hal Whitehead: "When you compare relative brain size, or levels of self-awareness, sociality, the importance of culture, cetaceans come out on most of these measures in the gap between chimps and humans. They fit the philosophical definition of personhood,"

which appeared in Angier, N. (June 26, 2010). Save a whale, save a soul, goes the cry. *New York Times*. Retrieved from http://www.nytimes.com/2010/06/27/weekinreview/ 27angier.html?pagewanted=all.

13. Keim, B. (July 19, 2012). New science emboldens long shot bid for dolphin, whale rights. *Wired Science*. Retrieved from http://www.wired.com/wiredscience/2012/07/cetacean-rights.
14. Dawkin, M. S. (June 8, 2012). Convincing the unconvinced that animal welfare matters. *Huffington Post*. Retrieved from http://www.huffingtonpost.com/marian-stamp-dawkins/animal-welfare_b_1581615.html.
15. Dawkins, M. S. (2012). *Why Animals Matter: Animal Consciousness, Animal Welfare, and Human Well-Being*. Oxford: Oxford University Press, 4.
16. Bekoff, M. (May 15, 2012). Dawkins' dangerous idea: we really don't know if animals are conscious. *Huffington Post*. Retrieved from http://www.huffingtonpost.com/marc-bekoff/animal-consciousness_b_1519000.html.
17. Wynne, C. D. L. (2004). The perils of anthropomorphism. *Nature, 428*, 606.
18. Goldacre, B. (2009). *Bad Science*. London: Harper Perennial, 100.
19. Francione, G. (June 4, 2005). Our hypocrisy. *New Scientist, 2502*, 51–2.
20. Rivas, E. (1997). Psychological complexity as a criterion in animal ethics. In M. Dol, S. Kasanmoentalib, S. Lijmbach, E. Rivas, & R. van den Bos (Eds.), *Animal Consciousness and Animal Ethics* (pp. 169–84). Assen: Van Gorcum.
21. See arguments in Chapter 7 of Dawkins, M. S. (2012). *Why Animals Matter: Animal Consciousness, Animal Welfare, and Human Well-Being*. Oxford: Oxford University Press.
22. Reiss, D. (2011). *The Dolphin in the Mirror: Exploring Dolphin Minds and Saving Dolphin Lives*. Boston, MA: Houghton Mifflin Harcourt, 248.
23. Delfour, F. (2010). Marine mammals enact individual worlds. *International Journal of Comparative Psychology, 23*, 792–810.
24. Dawkins, M. S. (2012). *Why Animals Matter: Animal Consciousness, Animal Welfare, and Human Well-Being*. Oxford: Oxford University Press.

INDEX